Measuring Academic Research

Measuring Academic Research

How to undertake a bibliometric study

ANA ANDRÉS

Chandos Publishing
Oxford • Cambridge • New Delhi

Chandos Publishing
TBAC Business Centre
Avenue 4
Station Lane
Witney
Oxford OX28 4BN
UK
Tel: +44 (0) 1993 848726
Email: info@chandospublishing.com
www.chandospublishing.com

Chandos Publishing is an imprint of Woodhead Publishing Limited

Woodhead Publishing Limited
Abington Hall
Granta Park
Great Abington
Cambridge CB21 6AH
UK
www.woodheadpublishing.com

First published in 2009

ISBN:
978 1 84334 528 2

British Library Cataloguing-in-Publication Data.
A catalogue record for this book is available from the British Library.

To Ana, José Luis, Gema and Sergi

Contents

List of figures and tables

Figures

Tables

About the author

Ana Andrés achieved her baccalaureate degree in Psychology in 2004 and her MS in Methodology for the Behavioural Sciences in 2006. Currently, she is an assistant lecturer at the Department of Methodology for the Behavioural Sciences at the Psychology Faculty of the University of Barcelona.

Her experience in teaching has been focused on psychometrics, specifically on the process of building, adaptation and validation of questionnaires. Her investigations centre on the application of methodological procedures to the Health Sciences and in this regard she has applied bibliometrics to scientific productivity in the field of obesity. She is also developing a questionnaire to assess processes of change for weight loss in overweight and obese people. A specific issue of interest in this research line is the adaptation of questionnaires that may be useful in the treatment of those who are overweight and obese as measures related to binge eating disorder.

The application of both bibliometric and psychometric procedures in the field of obesity is the topic of her doctoral thesis.

The author may be contacted via the publishers.

1

Introduction

This book aims to illustrate the main analyses applied in bibliometric studies. The objective of bibliometrics is basically to assess scientific literature in a given field, hence its broad applicability to all manner of disciplines. Consequently, readers can choose an analysis from the present book according to the characteristics of their own research field. The step-by-step explanations about how to carry out a bibliometric study, together with the many examples of the different indices and analyses that can be performed, provide a comprehensive guide to the main methods used in bibliometrics.

The book begins by defining *bibliometrics* and then considers the steps which must be taken before carrying out a bibliometric study. The following chapters focus on the specific analyses used to study the productivity of authors and journals. For example, we will see that it is possible to study collaborations between researchers, while the widespread use of citation analysis to evaluate scientific output will be considered with respect to both researchers and journals separately. The main limitations of the analyses shown will be discussed in order to give an overview of their uses and consequences in bibliometric research. Furthermore, it will be argued that a critical attitude should be adopted when choosing the most appropriate analysis for a bibliometric study.

In sum, the book offers a guide to the many analyses that could be included in a bibliometric analysis, and describes how to calculate and interpret them. Readers should take into consideration those tools which could be of interest for their bibliometric study, and seek to apply and interpret them as accurately as possible.

Bibliometrics: a science of science

The study of scientific literature has a long history dating back to the early decades of the past century. However, despite the amount of research in this area it was not until 1969 that the term *bibliometrics* first appeared in print (Pritchard, 1969). It was defined as the 'application of mathematical and statistical methods to books and other media of communication', and the term was quickly adopted and used, particularly in North America (Wilson, 1999). At almost the same time, Nalimov and Mulchenko (1969) coined the term *scientometrics* to refer to 'the application of quantitative methods which are dealing with the analysis of science viewed as an information process'. In contrast, this term was widely used in Europe (Wolfram, 2003). Initially, therefore, scientometrics was restricted to the measurement of science communication, whereas bibliometrics was designed to deal with more general information processes. At present, however, bibliometrics and scientometrics are used as synonyms (Glänzel, 2003).

In the 1990s the term *informetrics*, previously introduced by Gorkova (1988), was also used to designate a more general sub-field of information science that dealt with the statistical analysis of communication processes in science. Informetrics also deals with electronic media, including analyses carried out in electronic libraries (Glänzel, 2003). Finally, other terms such as *webometrics* or *cybermetrics* (Almind and Ingwersen, 1997; Björneborn and Ingwersen, 2001) have also been introduced to designate the study of scientific literature from electronic resources. In the present book the terms bibliometrics and scientometrics will be used without distinction to refer to the study of scientific literature.

The first evidence of bibliometrics dates back to 1873, when de Candolle described changes in the scientific strength of nations according to membership of scientific societies. With this study he aimed to identify factors that might influence the scientific success of a nation (van Raan, 2004). Subsequently, Lotka (1926) analysed the frequency distribution of scientific productivity, and his work led to the development of Lotka's law. This law, one of the most widely used in bibliometrics, assesses patterns in author productivity. Another pioneering study is that of Gross and Gross (1927) regarding citation analysis. These authors aimed to identify those journals with a high impact in their own research field, chemistry. This work has had enormous consequences, since citation analysis is now one of the main areas in bibliometrics. Another key study is that of Bradford (1934), who considered the frequency distribution of papers across journals. As in the case of Lotka's work, the resulting Bradford's law is now widely used in bibliometrics to study journal productivity. Another pioneering author was Zipf (1935, 1949), who studied the frequency of words in a text. His law can be considered as a generalisation of both Lotka's and Bradford's laws.

However, the real breakthrough in bibliometrics arrived some years later through the work of Garfield (1955) and Price (1963). Garfield developed a *Science Citation Index*, i.e. a multidisciplinary database in which authors could find articles from across many fields. This proved to be a visionary tool that greatly facilitated the researcher's task. The consequences of the indexation system proposed by Garfield will be widely discussed in this book. As van Raan (2004) states, his work has marked the rise of bibliometrics as a powerful field within the study of science. Another seminal work is Price's *Little Science, Big Science*, a book first published in 1963. This revolutionary book represented the first systematic approach to the structure of modern science that was applied to science as a whole. It also established the foundation of modern research evaluation techniques (Glänzel, 2003). The main contributions of Price's book to the development of scientometrics will be discussed below.

As the interest in bibliometric studies began to rise, specific publications started to appear. The first periodical publication in this area was the journal *Scientometrics*, founded by Tibor Braun in 1978. However, in the 1980s this interest in bibliometric studies came up against the limits facing researchers who wished to carry out this kind of study. The lack of availability of documents, the manual collection of data and the licence fees charged for obtaining documents all hindered progress in the field. The breakthrough came as a result of new technological developments during the 1990s (Glänzel, 2003) and the availability of online data regarding publications meant that the traditional indexation systems which compiled journal information in paper volumes were replaced by online databases.

There are now many specific or multidisciplinary databases providing indexation information for thousands of journals, papers, books and proceedings. Undoubtedly, the technological age has enabled enormous strides to be made in the field of bibliometrics.

Finally, it is worth mentioning the various applications that bibliometrics has at present. Table 1.1 shows the three bibliometric topics which Glänzel (2003) has identified as sub-areas of contemporary bibliometrics.

The bibliometric analyses discussed in the present book are considered mainly in terms of their practical application to scientific disciplines. Of course, bibliometric indicators can also have implications for science policy, and we will examine the assessment of different levels of productivity.

Table 1.1 Applications of bibliometric sub-areas

Sub-area	Application
Methodology research	These studies focus on the methodology used to carry out bibliometric research and refer to the development or improvement of bibliometric indicators. Researchers specialised in this area will basically be bibliometricians.
Scientific disciplines	These bibliometric studies may be conducted by researchers from any discipline. The aim is to apply bibliometric indicators to a given area of study. Consequently, these studies apply metrics in order to describe science.
Science policy	This is the most important topic in the field. Here, bibliometric studies are used to assess different levels of productivity. This research is conducted by policy-makers with the aim of deciding how to distribute available resources.

Little Science, Big Science

The behaviour of scientific productivity has been a traditional topic of study in scientometrics. By considering all the documents published in an area of research it is possible to determine how they are distributed according to different variables. This section discusses the main characteristics of scientific growth.

Derek J. de Solla Price was the first scientist to formulate a specific exponential growth law applied to science. It became his most famous contribution and is now known as Price's law. The law was presented and discussed in his most well-known publication, the book entitled *Little Science, Big Science* (Price, 1963). The term 'big science' refers to large-scale instruments and facilities, supported by funding from government or international agencies, in which research is conducted by teams or groups of scientists. In fact, this term was previously introduced by Weinberg (1961). Price's short book has had a huge impact on the formulation of scientific growth, as well as on the foundations of bibliometrics.

In his book Price explains how science has progressed from 'little science', which was traditionally carried out by a small group of erudite scholars who then became eminent in their field of study. In comparison, 'big science' is characterised by large amounts of money being invested in personnel and infrastructure. 'Big science' has now taken precedence over its forerunner, and investment in the advancement of science plays an important role in the economy of developed countries. However, the transition from 'little' to 'big science' has been gradual and less dramatic than it might seem at first sight. In order to analyse how this change from 'little' to 'big science' has come about, it is necessary to measure productivity over time. Price stated, on the basis of various numerical indicators taken from many fields and aspects of science, that there is regularity in the growth of its production. This growth pattern fits an exponential function. Consequently, science grows in a multiplicative way over time and, according to this exponential function, the growth rate will be proportional to the population size, i.e. the bigger the population is, the faster it grows.

Price's law has two main properties. The first is that its validity remains precisely constant across broad periods of time. Consequently, Price states that exponential growth in science has been maintained for two or three centuries. The second main characteristic of this exponential function applied to science is its rapid growth. Accordingly, the author states that the gross size of science in terms of personnel or publications tends to be duplicated in a time period of 10 to 15 years. In general terms, one could consider all the production achieved during this time, without taking its quality into account; as such, both high- and low-quality publications will be included in this count. If the requirement level is increased a little, one could consider that overall scientific production will be duplicated during a time period of 15 years. In this case, we would be more selective, including only good-quality authors and publications. Should we wish to be even more demanding and consider only very high-quality publications, then the necessary time period to duplicate production will increase to 20 years. The conclusion to be drawn from these statements is that science grows very rapidly and, consequently, the number of productions and scientists also grows in a multiplicative way.

Contemporary science

If production and scientists grow in a multiplicative way, then authors will be contemporary. In other words, if scientists are constantly duplicating every 15 years up to the present moment, a large proportion of scientists will be contemporary. Therefore, according to this rule, most science is current. This fact is a consequence of the accelerated growth. If, in 15 years' time, the number of scientists will have duplicated, then as many new scientists will appear during that time as existed in the whole of the previous period. However, scientists who are contemporaries in a given time period are not only new scientists who have appeared during that period. Consequently, Price (1963) estimated that approximately 87.5 per cent of all scientists there have ever been are currently living and, therefore, are contemporary. This value is known as the *contemporary coefficient*, and although it could be precisely calculated for a given point in time, Price stated that its value will always be close to 87.5 per cent due to the accelerated growth of science.

The 'end' of science and the obsolescence of literature

Given the *exponential growth* of publications and authors proposed by Price in his book, it would seem that science will never cease to grow. However, this assertion must be treated carefully, and it is necessary to analyse in detail the possibility that science will grow according to an exponential pattern. Firstly, we have to bear in mind the exponential function. Its profile is shown in Figure 1.1 and, applied to our subject, it will represent the growth in the number of publications or authors over time. As the function shows, the growing trend will always be increasing and, consequently, science never ceases to grow.

There is evidence that productivity can fit this pattern. One example is shown in the article by López-Muñoz et al. (2006), who carried out a bibliometric study of scientific publications related to bipolar disorder

Figure 1.1 Exponential function

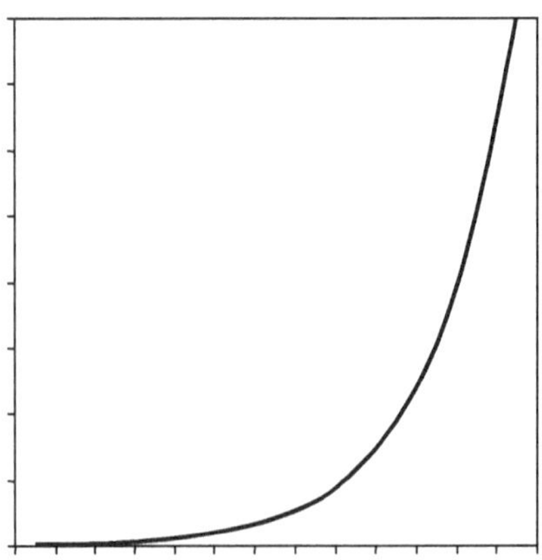

between 1980 and 2004. They identified over 4,000 articles on this topic and analysed their distribution over time; the frequency of studies over time was counted and represented graphically. However, in this article the authors go one step further and assess the nature of the growth profile in bipolar disorder research. Thus they calculated the fit of their data to both a linear and an exponential function. The results showed that the percentage of variance explained by the linear model was 73.67 per cent of data variability, while the exponential model explained 90.19 per cent of the variance. It therefore seems that scientific productivity in the field of bipolar disorder follows Price's law of exponential growth. Figure 1.2 shows the evolution of productivity over time as well as the linear and exponential functions for these data.

Although the exponential growth of productivity seems to be a valid explanation from a theoretical point of view, it becomes almost absurd when applying it to the real world. In real data, there is no undefined growth until infinity. On the contrary, exponential growth reaches a certain limit, after which the process becomes weaker rather than continuing to grow to an absurd extent. In fact, if the number of scientists were to grow exponentially, it would grow faster than the general population, which is nonsense.

Figure 1.2 **Growth of scientific literature in *bipolar disorder***

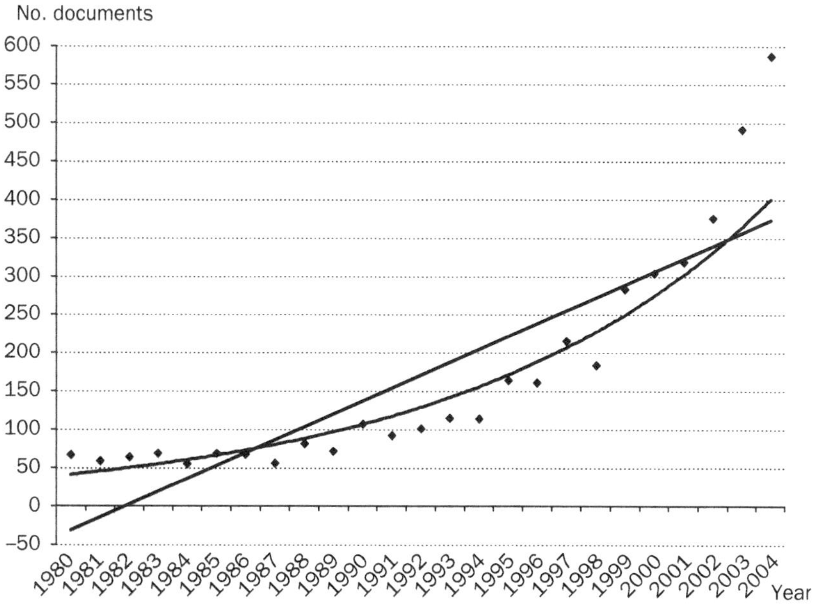

Reprinted with permission from López-Muñoz et al. (2006).

A more suitable explanation for scientific growth is thus to consider exponential growth as a phase in a more complex growth pattern. Thus accelerated growth may exist (although with fluctuations), but would be followed by the stabilisation of productivity. This perspective is more realistic and believable, and the pattern corresponds to the logistic curve. This function is represented by the shape shown in Figure 1.3.

The logistic curve is limited by a base that corresponds to the initial value of growth. This would mean that, at the beginning, no publications have been found. The curve also has a ceiling, after which growth is not possible in the usual way. As can be seen in the graphical representation, there is a phase in which growth is exponential but only up to a certain value, after which growth becomes stabilised.

Consequently, it can be assumed that the exponential growth described by Price's law is acceptable within a logistic function, so this period of accelerated growth will be followed by a stabilisation phase. In the example described above (López-Muñoz et al., 2006), we can conclude that the exponential growth of productivity in the field of bipolar disorder is only the beginning of a logistic curve.

Within this context, the term 'big science' corresponds to this phase of exponential growth. However, it is not possible to know when this increase in productivity will reach the point of stabilisation. In any event, this will never mean the end, but simply the progression to a new phase.

Related to the ideas already discussed regarding the growth of science, it is also worth mentioning the concept of the *obsolescence of literature*. The term obsolescence of literature was introduced by Price (1963) and

Figure 1.3 **Logistic curve**

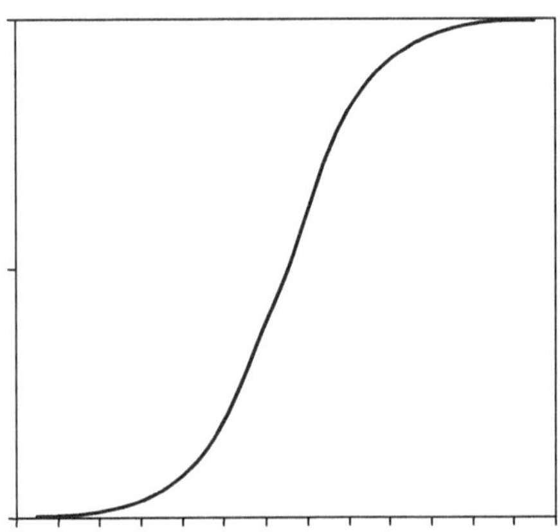

refers to the decline in the use of documents over time. This concept is closely related to citation analysis, as the use of a document is measured by the number of citations it receives. While the productivity of a given field is advancing, the citations received for a given document usually decline over time. Finally, if the document is no longer cited it will become obsolete. Only a small group of articles will remain frequently cited over time, and these will become classics due to the fact that authors continue to cite them in recognition of their contribution. However, very few documents will achieve the status of classic articles (Wolfram, 2003).

A closer look at bibliometric analysis

As will be illustrated through the different examples included in this book, scientific productivity can be analysed in any research field, whether it be from natural science or the social sciences and humanities. The only requirement is to gather a set of publications about a given field. The present section details the first steps in conducting a bibliometric study.

It is important to begin with a clear *topic definition*. If the aim is to describe productivity in a given area, we must be sure that the documents included in our study are truly representative of the research field. As such, the documents gathered have to cover the whole domain of the field being analysed.

It should be noted that the topic definition will depend on the field we are studying, and therefore it will influence both the documents to be gathered and the analysis to be carried out.

Although a bibliometric study can be applied to define general productivity in a given area it may also be used to evaluate the productivity of individual researchers, journals, countries or any other levels of performance. Therefore it is also necessary to define at the outset the kind of data which will be assessed. This decision will determine whether we calculate several or just one of the many indices that will be presented in this book. For instance, if we are assessing a researcher's performance, one or two indices may be enough to provide an idea about the productivity and quality of this scientist. However, if the aim is to describe in detail the current state of a particular field it may be appropriate to use a wider range of indicators. As such, bibliometric indicators must be carefully selected depending on the aim of our study.

Having selected the level of productivity to be assessed (authors, research groups, journals, countries, a particular field, etc.) the next step is to conduct a *bibliographic search* in order to collect the representative documents for our study. Since the documents selected and their content form the basis of our analysis, the bibliographic search is a key stage in this process. Although the documents to be analysed in a bibliometric study may be obtained from a specific database that collates publications about a given area of study, it is also possible to search for documents in a multidisciplinary database, thus gathering publications from across a range of disciplines. This book makes particular reference to one of the most widely used multidisciplinary databases, the *ISI Web of Science* (*WoS*), although others are also available, for example *Scopus* or *Google Scholar*. These databases are discussed in the present book due to their impact and usefulness to the scientific community.

The characteristics of a database that need to be considered when deciding whether it is appropriate for a given bibliometric study have been discussed by Neuhaus and Daniel (2008). They point out that databases contribute in two ways to the development of a bibliometric study: firstly as a data source, but also as a platform that provides the analytical tools necessary for the bibliometric calculations (Hood and Wilson, 2003).

The first – and perhaps most important – characteristic they identify is *coverage*. This is the extent to which the sources processed by the database cover the written scholarly literature and, in particular, the journal literature in a field. Obviously, it is essential that the documents on which a bibliometric study is based actually cover the content we aim to analyse. Thus we must ensure that coverage is not biased towards certain countries, languages, publishers or types of documents. Furthermore, depending on the area of research we are working in, we might be more interested in gathering papers from journals, books, dissertations, etc., or documents in languages other than English.

Another characteristic identified by these authors is the *consistency and accuracy of data*. Databases inevitably contain inconsistent and erroneous data, and therefore it is important to select one that minimises these errors. For example, author names or surnames are sometimes written in an inconsistent manner, making it difficult to identify them, while institutional affiliations or journal titles may also include linguistic variations that can bias our data. In fact, institutional affiliations represent a broad source of bias due to the difficulty of clearly identifying author affiliations; at times only the department, faculty or university of an author is listed, making it difficult to standardise the format. Additionally, hospitals may be associated with university colleges, and some research groups prefer to be affiliated to the group's name. Consequently, a database that minimises these inconsistencies will also reduce the amount of manual checking required to ensure the reliability of the data.

Another characteristic that has to be taken into consideration when selecting a database is the *data fields* required. This means that prior to carrying out a bibliometric study we need to have an idea about the analysis which will be performed. For example, the information required to study the contribution of a single scientist will be different from the information needed to study institutional productivity. Thus when carrying out a bibliographic search it is important to select all the required data fields. The data fields which we may be interested in selecting include variables such as author names, institutions, number of citations received, year of publication, year of the citations received, number of authors contributing to the publication and subject category to which the journal belongs. Alternatively, interest may lie in studying certain types of documents such as research articles, reviews, letters or comments, or different types of publications like journal articles, books or proceedings. Of course, the data fields in which we are interested will also influence the database chosen for the bibliographic search, as not all databases may provide the desired information.

The *browsing options* of a database are another characteristic discussed by Neuhaus and Daniel (2008) and this is closely related to data accuracy. One of the analyses that can easily be influenced by data errors is citation analysis, which is based on the citations received by an author or document. Ideally, the database provides a match of those publications that have cited a given author or paper, but if the citing source has made a mistake the match will not be done correctly. It is therefore important that databases include browsing options so as to identify these possible inconsistencies.

As mentioned above with respect to the data fields of interest, we have to be sure that the database being used actually provides the information we are looking for. This will depend on the *search options*, another characteristic of a database. Each database has an interface, with different search, browsing and saving options. Depending on our objectives, we might be interested in saving all the author names on a paper or the publications that cite a given author, or in ranking papers according to the number of times they have been cited. The availability of these options will depend on the database chosen.

Furthermore, there are also databases whose *analytical tools* are capable of analysing the results obtained in a bibliographic search. These can provide information about the most productive authors among the publications identified or indicate the rate of productivity over time. However, the analytical tools of databases can only manage a certain number of search results, so depending on the number of documents in our bibliographic search it is possible that a given database will be unable to analyse all these records at the same time. Moreover, there are some databases which include the option of analysing collaboration maps between authors. All in all, this illustrates that there are many characteristics which can influence the decision to choose one or another database.

The final characteristic pointed out by Neuhaus and Daniel (2008) is the *saving and exporting options*. The last step in a bibliographic search consists of saving the documents found in a file that will then be used to conduct the bibliometric analysis. Some databases offer different formats for saving records or exporting them to bibliographic software. Once again, we have to weigh up our different needs in order to decide which database to use.

One final consideration before moving on to look at the main analyses used in bibliometric studies is the importance of a *clear explanation of the bibliographic search* conducted. Since our analysis will lead to a bibliometric report, we must ensure that the procedure used to obtain the target documents is well defined. This means explicitly stating the period of time that has been taken into consideration, the database or databases used, and the keywords entered. Any other procedure used to select documents should also be specified (for example, a manual check of document content, exclusion criteria, etc.). This systematic procedure will enable other authors to replicate our study.

2

Descriptive analyses

The easiest way to begin a bibliometric study is through descriptive analyses of the main characteristics of the studies that will be included. These will offer a quick and even visual impression of certain aspects related to productivity in the given field of study.

Temporal evolution

Once all the documents that will form part of the study have been collected it is then possible to analyse how they have evolved over time. In order to conduct this analysis it is necessary to know the year of publication of each study and to calculate the frequency of studies published each year. It will then be possible to identify the trend of scientific productivity in a given area of study. All the significant information about temporal evolution, such as year, number of studies (frequency), percentage and cumulative percentage is usually collated in a table. These data, along with a graphical representation, will show the trend in output.

A good example of the temporal evolution of scientific productivity can be seen in the area of the *human immunodeficiency virus* (HIV), which leads to acquired immunodeficiency syndrome (AIDS). Table 2.1 shows, over the years, the number of publications found in the Medline database containing the acronym HIV in the topic.

Although a couple of articles about immune deficiency diseases were published during the 1970s, it was not until the 1980s that a clear interest in this topic emerged. From this point on, one can say that an information explosion took place. This result is consistent with the fact that it was in 1983 that the human immunodeficiency virus was found to lead to AIDS. A graphical representation of the temporal evolution of HIV research (Figure 2.1) enables us, at a glance, to understand the data shown in

Table 2.1 Publications in Medline about *HIV*

Year	Number of publications	%	Cumulative %
1970	1	< .001	< .001
1975	1	< .001	< .001
1982	29	0.01	0.01
1983	696	0.33	0.34
1984	1,153	0.54	0.89
1985	1,759	0.83	1.71
1986	2,944	1.39	3.10
1987	4,649	2.19	5.29
1988	7,433	3.50	8.79
1989	7,627	3.59	12.38
1990	8,531	4.02	16.39
1991	9,277	4.37	20.76
1992	9,546	4.49	25.25
1993	9,543	4.49	29.75
1994	9,275	4.37	34.11
1995	10,241	4.82	38.93
1996	10,568	4.97	43.91
1997	10,331	4.86	48.77
1998	10,999	5.18	53.95
1999	10,638	5.01	58.96
2000	10,190	4.80	63.76
2001	10,366	4.88	68.64
2002	10,173	4.79	73.42
2003	10,571	4.98	78.40
2004	10,987	5.17	83.57
2005	11,352	5.35	88.92
2006	11,514	5.42	94.34
2007	12,019	5.66	100.00

Table 2.1. Figure 2.1 clearly shows a sharp increase in research activity up until 1990, followed by a constant increase up to the present day.

Another example of the temporal evolution of productivity is that provided by Miguel-Dasit et al. (2008). These authors analysed how

Figure 2.1 Temporal evolution of scientific productivity about *HIV* research in Medline

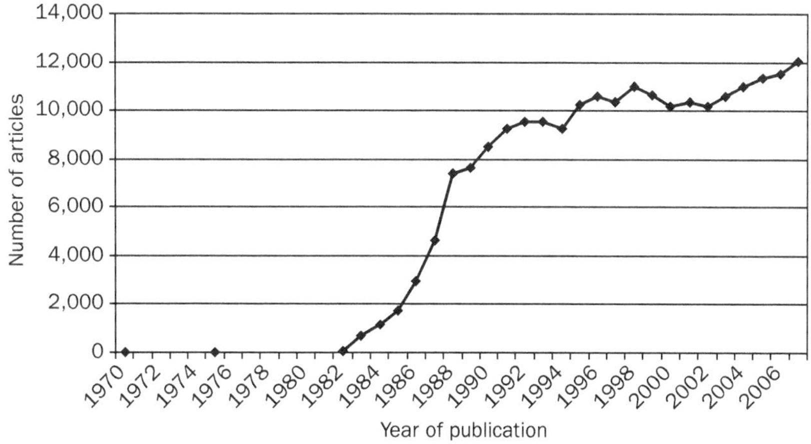

productivity in the field of *magnetic resonance* (for Spanish researchers) progressed from 2001 to 2007 in the Medline database. As can be seen in Table 2.2 they divided the articles found into different categories according to topics within magnetic resonance research. They also provide a general count of the number of papers published during this period.

Depending on the number of studies, it may also be possible to represent the temporal evolution graphically in terms of ranges of years rather than specific years.

The temporal evolution of productivity can, in fact, be consulted via online databases. One database that offers this resource is the ISI Web of Knowledge. Thus when performing a search in any of the ISI Web of Knowledge databases it is possible to obtain information about the number of publications in a given period of time by means of a single click. When a search is carried out of a given topic, a list of articles is shown, along with the option 'Analyse Results'. Selecting this option yields the number of publications in a field according to certain criteria, such as the year of publication.

The same analysis regarding temporal evolution could, alternatively, focus on productivity over time in a given journal. In this case, the number of publications will refer to those published in a specific journal.

By way of example Table 2.3 considers the report by Kirchler and Hölzl (2006), who conducted a bibliometric study of the *Journal of Economic Psychology*. It can be seen that the temporal evolution of

Table 2.2 Temporal evolution of productivity in *magnetic resonance* production in Spain from 2001 to 2007

Topic	No. of papers published in 2001	No. of papers published in 2002	No. of papers published in 2003	No. of papers published in 2004	No. of papers published in 2005	No. of papers published in 2006	No. of papers published in 2007	Total no. of papers per topic
Abdominal	4 (17%)	1 (3%)	7 (27%)	3 (7%)	7 (17%)	5 (17%)	2 (3%)	29 (100%)
Breast	–	1 (11%)	–	2 (22%)	3 (33%)	2 (22%)	1 (11%)	9 (100%)
Cardiac	1 (5%)	3 (14%)	3 (14%)	1 (5%)	6 (28%)	4 (19%)	3 (14%)	21 (100%)
Chest	–	1 (25%)	–	–	–	3 (75%)	–	4 (100%)
Contrast media	1 (17%)	1 (17%)	3 (49%)	1 (17%)	–	–	–	6 (100%)
Genitourinary	5 (31%)	2 (6%)	1 (6%)	4 (19%)	2 (12%)	3 (19%)	1 (6%)	17 (100%)
Head and neck	3 (75%)	–	–	1 (25%)	–	–	–	4 (100%)
Musculoskeletal	8 (13%)	10 (12%)	12 (20%)	11 (16%)	8 (14%)	8 (14%)	10 (13%)	67 (100%)
Neuroradiology	18 (13%)	27 (19%)	15 (11%)	18 (13%)	14 (9%)	25 (19%)	22 (14%)	139 (100%)
Pediatrics	–	2 (22%)	2 (22%)	2 (22%)	2 (22%)	1 (11%)	2 (11%)	11 (100%)
Vascular	1 (12%)	2 (12%)	–	4 (37%)	1 (12%)	2 (25%)	–	10 (100%)
Computer applications	–	–	–	1 (33%)	–	1 (33%)	1 (34%)	3 (100%)
Miscellaneous	–	–	3 (27%)	2 (18%)	1 (9%)	3 (27%)	2 (18%)	11 (100%)
Total no. of papers and mean IF of MR publications	41 (12%)	50 (15%)	47 (14%)	51 (15%)	44 (13%)	57 (17%)	42 (12%)	332 (100%)

Reprinted with permission from Miguel-Dasit et al. (2008).

Table 2.3 **Temporal evolution of publications in the *Journal of Economic Psychology***

	1981-85	1986-90	1991-95	1996-2000	2001-05	Total
Articles published	127	135	182	190	220	854
Page count, 5% trimmed mean	16.47	18.43	18.26	18.73	17.51	17.92
Reference count, 5% trimmed mean	23.22	30.97	29.64	29.68	38.35	30.99
Author count, 5% trimmed mean	1.59	1.70	1.71	1.77	2.02	1.77
Percentage of single-author articles	52.0%	43.7%	41.2%	39.5%	34.1%	41.0%

Reprinted with permission from Kirchler and Hölzl (2006).

productivity in this journal is analysed in terms of the number of articles published over time. Other descriptive data regarding certain characteristics of the articles are also included in this table.

Number of authors

Another descriptive analysis which might be of interest as part of a bibliometric study concerns the number of authors contributing to each publication.

A count of the number of authors who have contributed to reports offers some indication of the degree of collaboration between authors. An example of such an analysis is the paper by Davarpanah and Aslekia (2008), who report that just over half the articles included in their study (51.1 per cent) had a single author, while the other half (48.9 per cent) had two or more contributing authors. Alternatively, an illustration of the authorship pattern could replace this report of percentages (Figure 2.2), as in the article by Andrés et al. (2007).

Another descriptive analysis commonly used in bibliometric studies concerns the most productive authors. This type of analysis draws up a list of the most productive authors, as well as the total number of publications. The percentage of publications by these authors with respect to the total number of publications included in the study or over

Figure 2.2 Illustration of the number of authors contributing to the publications

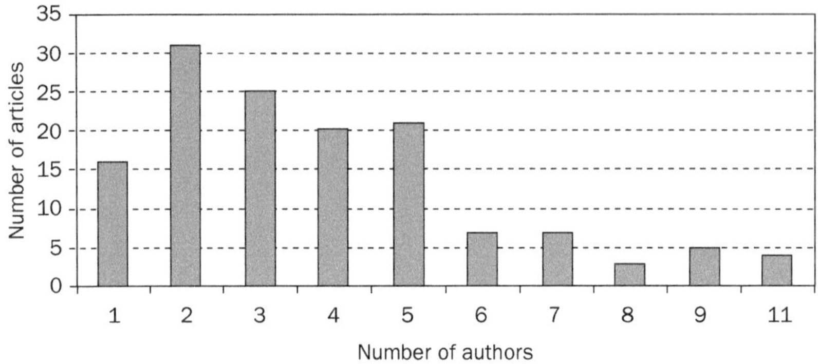

Reprinted with permission from Andrés et al. (2007).

the period of time in which they have been most productive can also be included in this descriptive analysis.

These analyses regarding authors are complementary and the choice of which to use will depend on their suitability for describing the data at hand and the characteristics of each bibliometric study.

More detailed information about collaboration between authors will be provided in Chapter 5.

Institutions and countries

Another common analysis involves drawing up a list of the most productive institutions or countries in a given field. In this regard, it is usual to include a ranking of those who have collaborated in the greatest number of publications included. The procedure used to identify the most productive institutions or countries consists of obtaining the affiliation of the authors who have taken part in writing the documents. In order to get a more accurate analysis it is advisable to list the affiliations of all the authors rather than only that of the first author. Although this is a more laborious task it provides a better description of author participation as well as of collaboration between institutions and countries.

A widely used type of list is that concerning production percentages of the most productive institutions. An example of this approach is the paper by Willett (2007), who conducted a bibliometric study of productivity in the *Journal of Molecular Graphics and Modelling*. In this report, although a total of 687 institutions were associated with the authors included in the study, only those which provided at least 1 per cent of the journal's articles were listed (Table 2.4).

It should be noted at this point that the analysis of scientific productivity by institution faces an important technical problem to do with how the affiliation information of authors is obtained and arranged. Firstly, and as van Raan (2005) pointed out, the same institution may be referred to by a different wording. This author found, for instance, that Leiden University had up to five different variants of its name in the same database: Leiden University, Universiteir Leiden, Leiden Observatory, Leiden University Medical Center and Leids Universitair Medisch Centrum. As a result, it is sometimes difficult to distinguish the many institutes within the main organisation. Furthermore, due to language differences the same institution could be mentioned in more than one language. In order to overcome this problem, researchers need to be extremely rigorous when refining their database prior to carrying out a bibliometric study. This preliminary and laborious work is the best way of ensuring that it will not be necessary to repeat the analysis.

Another analysis regarding institutions involves the distinction between types of institution. Although less widely used, this approach counts how many authors were affiliated to public or private institutions, or to academic or non-academic ones.

Table 2.4 **Institutions providing at least 1% of the articles published in the *Journal of Molecular Graphics and Modelling***

Institution	%
University of Sheffield	2.2
University of Oxford	2.0
University of California at San Francisco	1.9
CNRS	1.8
University of Cambridge	1.5
University of North Carolina	1.5
Scripps Research Institute	1.3
University of Minnesota	1.2
Birkbeck College, University of London	1.0
Université de Paris 07	1.0
Naval Research Laboratory, Washington	1.0

Reprinted with permission from Willett (2007).

The paper by Willett (2007) also shows a list of the most productive countries, once again in terms of percentages. In this case, a total of 55 countries were involved in the publication of the articles included in the study. However, only those which provided at least 2 per cent of the articles published in the journal were shown (Table 2.5).

Table 2.5 **Countries providing at least 2% of the articles published in the *Journal of Molecular Graphics and Modelling***

Country	%
USA	38.1
England	18.3
Japan	7.2
France	6.8
Germany	4.3
Australia	3.5
Spain	3.5
Switzerland	3.0
Canada	2.7
Italy	2.7
Sweden	2.7
People's Republic of China	2.3

Reprinted with permission from Willett (2007).

Other indicators

Other descriptive analyses of the characteristics of research papers included in a bibliometric study may concern the language of the document, the type of literature or the subject category to which the document belongs. Thus it is possible to count how many of the publications included are written in each one of the *languages* found. As most of the journals listed in ISI and the most widely used databases are published in English, it is inevitable that this will be the main language used when publishing. Nonetheless, studies in other languages may also be found. For example, Chiu and Ho (2007) conducted a search for *tsunami research* in the Science Citation Index (SCI) of the ISI Web of Science and found that although most of the studies were published in English

(95.0 per cent), some of them were written in Russian (4.2 per cent), French (0.7 per cent) or Spanish (0.1 per cent).

The *type of document* was also a topic of study for these authors, since they counted how many documents belonged to each one of the literature types. They found that the most frequent type of document was the article (88.0 per cent), followed by reviews (4.9 per cent), editorial materials (2.9 per cent), new items (1.9 per cent), notes (1.1 per cent), letters (0.5 per cent), corrections (0.2 per cent), meeting abstracts (0.2 per cent), biographical items (0.1 per cent), book reviews (0.1 per cent) and corrections/additions (0.1 per cent).

Finally, some studies include a distribution of articles across different and previously established *subject categories*. The ISI Web of Knowledge has 56 fixed subject categories for the database of the Journal Citation Reports (JCR) Social Sciences Edition and 173 subject categories for the JCR Science Edition. Thus each journal listed in these databases will be associated with one or more subject categories.

An example of a subject category analysis is shown in Table 2.6, where Chiu and Ho (2007) divided up the publications found in the ISI subject categories. Once again, only the most representative categories were shown.

However, other subject categories could be used in the analysis, as did Davarpanah and Aslekia (2008) when they categorised documents according to the Library and Information Science Abstract (LISA) Board Subject Headings.

Table 2.6 **ISI subject categories with most of the publications**

Ranking	Subject category	P	% P
1	Geosciences, Multidisciplinary	348	26.0
2	Geochemistry & Geophysics	222	16.0
3	Oceanography	115	8.5
4	Geology	95	7.0
5	Water Resources	78	5.7
6	Meteorology & Atmospheric Sciences	70	5.1
7	Geography, Physical	68	5.0
8	Multidisciplinary Sciences	40	2.9
9	Engineering, Civil	29	2.1
9	Engineering, Ocean	29	2.1

Reprinted with permission from Chiu and Ho (2007).

3

Author production

Although basic descriptive analyses regarding the most productive authors can be carried out to identify the most highly productive people in a given area of research, the data can also be treated in another way. Indeed, author productivity is usually analysed according to a widely used bibliometric law: Lotka's law. The aim of this famous law and how it should be applied to data will be discussed below. An analysis based on Lotka's law provides a more specific interpretation of author productivity.

Lotka's law

Alfred J. Lotka (1926) studied author productivity patterns and developed one of the main laws in bibliometrics. He observed that, in a given area of science, there are a lot of authors who publish only one study, while a small group of prolific authors contribute with a large number of publications. This premise is the basis of Lotka's law, also known as the *inverse square law* on author productivity. The law takes the number of authors who have contributed with a single study and then predicts how many authors would have published x studies, according to this inverse square law. In summary, the number of authors who produce x studies is proportional to $1/x^2$.

Let us suppose that, in a given field, 100 authors have published a single study. We can then predict how many authors would have published twice, according to the following formula: $y_x = c \times 1/x^2$ (although more commonly used is its reformulation $y_x = c \times x^{-2}$). In this formula, y_x is the number of authors with x publications, c is the number of authors with a single publication (100 in this hypothetical case) and x is the number of publications itself. Therefore, if we want to know how many authors have contributed with two articles ($x = 2$) we find that $y_2 = 100/2^2 = 25$. Thus it is predicted that 25 authors would have published two articles. As stated above, the number of prolific authors will decrease in a progressive way, so only 11 authors would have published three articles according to this formula ($y_3 = 100/3^2 = 11$). As the number of publications increases, the number of authors who have published x articles decreases. According to these calculations, 25 per cent of contributions will correspond to 75 per cent of the – less productive – authors, whereas the most productive authors (only about ten) will account for 50 per cent of contributions.

This author productivity pattern does not seem to depend on the science in which Lotka's law is applied. The only condition that needs to be assumed is the period of time. If we expect to find a small group of prolific authors it is necessary to consider a long enough period of time so that they have the opportunity to publish more than once or twice. This period is set at ten years or more.

According to the distribution described above, an author production profile is not randomly distributed. This means that, up to a point, the more articles produced by an author the more likely he or she is to produce others. Therefore productivity is related not to the number of articles published by an author but to its logarithm. As such, it appears to become progressively easier to publish once an author has published a large number of articles. This fact could be described as a *cumulative advantage* which is described by Merton (1968) as the *Matthew effect*, i.e. the phenomenon by which the rich get richer while the poor get poorer. Applied to author productivity, those who have more publications are also more likely to obtain other results (proportional to the publications previously obtained). Consequently their productivity will grow in an accelerated way.

Lotka's law has been commonly applied in scientometrics and author productivity from many fields has been shown to fit the distribution it proposes.

How to apply Lotka's law

The first step in applying Lotka's law in order to know if author productivity fits with this pattern is to decide which data will be considered for analysis. As Wolfram (2003) stated, publications with multiple authors can present a dilemma in productivity studies because there are several ways for authors to receive credit for publications they have co-authored, as shown in Table 3.1.

Of the options shown in Table 3.1, the complete count process has received the greatest support from the literature (Egghe and Rousseau, 1990; Lindsey, 1980; Rousseau, 1992).

Table 3.1 **Author's credit for publications**

Author's credit	Description
Complete count	Each occurrence of an author is recognised and receives equal treatment, regardless of the number of authors associated with the article.
Straight count	Only the first author is counted, based on the assumption that the first author is the primary contributor to a publication.
Adjusted count	Authors receive fractional credit for publications with multiple authors. According to this adjusted count, each author within a publication with two authors will receive 0.5; 0.2 for five authors, etc. The credit received also may be weighted depending on the number of authors, so the preceding authors will receive more credit than the subsequent authors.

Although the general formulation of Lotka's law assigns an a priori value to the n exponent ($n = 2$), it can be calculated for each author productivity distribution. Exponent n can vary and may be higher or lower than 2. In fact, when the distribution includes highly prolific authors, the difference between the number of high-output and small-output authors will be larger. In this case, the number of highly prolific authors will decrease faster than the inverse square, and the exponent n will be closer to 3.

The methods for calculating the specific coefficients when applying Lotka's law to a given distribution have been defined (Nicholls, 1986; Pao, 1985).

In order to understand how to apply Lotka's law step by step, we will take the data from a hypothetical example. Let us imagine that a bibliographic search has been carried out in a given field for a given period of time. As stated previously, it is advisable to use the complete count, so we will consider all the authors contributing to the articles. Table 3.2 shows the number of articles contributed to by each author.

In this example, authors have contributed between one and ten articles. However, as expected, most authors have contributed with a small number of articles, while a small group of authors has been very prolific.

Lotka's law can now be applied to test whether these data fit the law. To this end, the data must be gathered in a new table that includes the additional information needed for Lotka's calculation (Table 3.3).

Table 3.2 **Example of documents count considering all authors in a determinate field**

Number of works	Number of authors
1	1,005
2	130
3	32
4	15
5	7
6	6
7	4
8	3
9	1
10	2

Let us consider this table step by step. The first two columns of Table 3.3 correspond to the data shown in the previous Table 3.2, indicating the number of contributions made by each author. Specifically, column x corresponds to the number of studies published by the authors in this field. In this case, authors have contributed between one and ten articles. The next column, y_x, corresponds to the number of authors publishing a given number of articles. It can be seen that 1,005 researchers in this area have contributed a single article, while two authors have been the most prolific, contributing ten articles each.

As is shown, a total of 1,205 authors ($\sum y_x$) have been involved in producing the articles included in this bibliometric study. The next two columns (X and Y) correspond to the logarithm of the frequency of articles (x) and authors (y_x), and the product of these is shown in the next two columns. As the totals for these data are also available at the bottom of Table 3.3, it is possible to calculate the frequency of authors with a single article, and those with two, three, etc. (corresponding to column $y_x/\sum y_x$).

So far, these data have been directly obtained from the frequency of authors publishing x articles. Once this frequency (the observed frequency) has been obtained, Lotka's law can be applied to obtain the expected frequency of authors publishing x articles. Thus it is necessary to calculate the n exponent for this particular case.

Although Lotka's law proposes a growth in production according to its formula $y_x = c \times x^{-n}$, where n is equal to 2, the law has to be tested for the data in question. This will yield the exponent n, which corresponds

Table 3.3 Lotka's law

x	y_x	$X = \lg x$	$Y = \lg y$	X^2	XY	$y_x/\sum y_x$	$\sum(y_x/\sum y_x)$	f_e	$\sum f_e$	D
1	1,005	0.000	3.002	0.000	0.000	0.834	0.834	0.812	0.812	0.022
2	130	0.301	2.114	0.091	0.636	0.108	0.942	0.114	0.926	0.016
3	32	0.477	1.505	0.228	0.718	0.027	0.968	0.036	0.962	0.006
4	15	0.602	1.176	0.362	0.708	0.012	0.981	0.016	0.978	0.003
5	7	0.699	0.845	0.489	0.591	0.006	0.987	0.009	0.987	0.000
6	6	0.778	0.778	0.606	0.606	0.005	0.992	0.005	0.992	0.000
7	4	0.845	0.602	0.714	0.509	0.003	0.995	0.003	0.995	0.000
8	3	0.903	0.477	0.816	0.431	0.002	0.998	0.002	0.997	0.001
9	1	0.954	0.000	0.911	0.000	0.001	0.998	0.002	0.999	0.001
10	2	1.000	0.301	1.000	0.301	0.002	1.000	0.001	1.000	0.000
	1,205	6.560	10.801	5.215	4.500					

to the present distribution of author productivity and also indicates whether the data really fit Lotka's law.

Consequently, the first step is to calculate the exponent *n* using the least squares method and according to the following formula:

$$n = \frac{N\sum XY - \sum X \sum Y}{N\sum X^2 - (\sum X)^2}$$

All the data needed for the *n* formula can be obtained from Table 3.3. The only index that requires further work is *N*, which represents the number of pairs considered. In this example, those authors who have published between one and ten articles will be considered, so this will represent ten pairs of data ($N = 10$).

There is one specific case where not all the pairs of data are included in the analysis of Lotka's law. This is when $y_x = 1$ is found at the end of the distribution, corresponding to the highest values of x (number of articles). In such cases, this small group of most prolific authors are excluded from the analysis in order not to overestimate the results. However, these pairs of data have to be carefully excluded. If $y_x = 1$ is located not at the end of the distribution but within other pairs of data, then it should be included in Lotka's calculation.

An example of this is shown in Pulgarín and Gil-Leiva (2004), where those pairs of data in which $y_x = 1$ were excluded from the analysis as they were located at the bottom end of the distribution. In this article, Lotka's law was calculated based only on the authors who had published up to seven articles (Table 3.4).

In the example provided in this chapter, none of the data pairs were excluded. This is because $y_x = 1$ was not found to be associated with the highest value of x.

In our specific example, the value of *n* is obtained from the data in Table 3.3 and assuming that $N = 10$. Inserting the corresponding values into the *n* formula stated above gives:

$$n = \frac{10 \times 4.5 - (6.56 \times 10.801)}{10 \times 5.215 - (6.56)^2} = -2.83$$

Thus the value of *n* (absolute value) is 2.83, which will then be the specific value of the coefficient in Lotka's formula that will explain author productivity in this particular case.

Table 3.4 Pairs of data exclusion

x	y_x	
1	703	
2	112	
3	35	
4	11	
5	8	
6	7	
7	3	
9	1	Excluded data
10	1	Excluded data
14	1	Excluded data
34	1	Excluded data
Σ	883	

Reprinted with permission from Pulgarín and Gil-Leiva (2004).

As the formula of Lotka's law is $y_x = c \times x^{-n}$ and n is now known, the only index left to calculate is c. This value is obtained as follows:

$$c = \frac{1}{\sum 1/x^n}$$

For our data, c will take the value 0.81, and thus Lotka's formula will be $y_x = 0.81 \times x^{2.83}$. Here, y_x represents the expected frequency of authors publishing x documents. In order to avoid confusion in the nomenclature, let us define the expected frequency of Lotka's law as f_e, to distinguish it from the observed frequency (y_x). By introducing the values taken by the number of articles variable (x) it is now possible to obtain the corresponding expected frequencies. The values taken by f_e are shown in Table 3.3, along with its cumulative frequency ($\sum f_e$).

As the aim of this analysis is to determine whether these data fit Lotka's law, it is necessary to know the magnitude of the difference between observed and expected frequencies. This difference is shown in the last column of Table 3.3 and is computed by subtracting the cumulative expected frequency from the cumulative observed frequency: $\sum(y_x/\sum y_x) - \sum f_e$. This difference, in its absolute value, is shown in the D column.

Finally, the Kolmogorov-Smirnov test is applied in order to verify whether the observed data fit the theoretical distribution according to Lotka's law. The highest value in column (D_{max}) is taken as reference for comparison with the critical value (c.v.) whose general formulation is:

$$\text{c.v.} = \frac{1.63}{\left(\sum y + \left(\sum y/10\right)^{1/2}\right)^{1/2}}$$

In this example, the critical value is 0.047, obtained from the following formula:

$$\text{c.v.} = \frac{1.63}{\left(1,205 + \left(1,205/10\right)^{1/2}\right)^{1/2}} = 0.047$$

Since the maximum difference (D_{max}) obtained from Table 3.3. is 0.022, which is smaller than the critical value (0.047), the null hypothesis has to be accepted. We can therefore conclude that author productivity in this hypothetical research area fits Lotka's law.

4

Journal productivity

Another common unit of analysis in a bibliometric study concerns the journals in which the articles gathered are published. As in the case of author productivity, descriptive analyses about the most productive journals can be carried out. However, a more detailed analysis about the scatter of journal productivity may be interesting for research purposes. This chapter considers the main law applied to journal productivity, namely Bradford's law, and provides a detailed explanation of its application.

Bradford's law

Bradford's law was formulated in 1934 by Samuel C. Bradford with the aim of studying the distribution of scientific literature. His work was developed in the area of geophysics between 1931 and 1933, during which time he gathered all the articles he could find related to this area. Upon analysing the journals in which these articles were published he found a regularity, namely an *inverse relationship* between the number of articles published in a subject area and the number of journals in which the articles appear. This means that, in a given subject area, a small number of journals account for a sizeable portion of the total publications in that area, whereas increasing numbers of journals publish fewer articles in that area.

This law about journal productivity can easily be represented in graphical form (Figure 4.1). Journals are ranked and divided into groups or categories, depending on the number of articles they account for. These groups are termed Bradford zones. The criterion for establishing these zones is that the number of articles in each zone has to be the same. However, the number of journals publishing these articles will not be the same in each zone, as some journals will be more productive than others.

Figure 4.1 **Bradford zones**

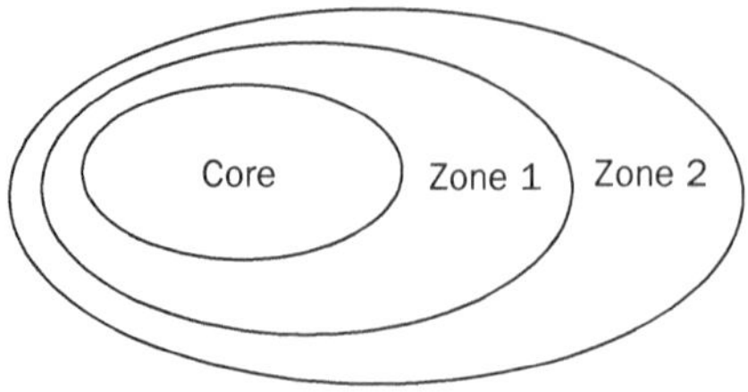

Given that Bradford's law ranks journals according to their productivity, a small group of articles will be located in the first central zone, while an increasing number of journals will be found in each subsequent zone. The first group comprises the *core* journals and will contain a given number of articles. While the number of articles will remain constant in all zones, the number of journals will increase across the zones. The ratio between the number of journals in subsequent zones has been observed to be approximately $1{:}n{:}n^2$ ...

This means that, for a given subject area, we must gather all the articles published in a given period of time and list the journals in which they have been published. Let us suppose that we have gathered a total of 300 articles published in 65 journals. If we accept the hypothesis that our subject area will fit a three-zone Bradford law there will be approximately 100 articles in each zone. The task is then to rank the journals according to their productivity and to count how many journals belong to the first zone (most productive journals). Suppose that the first zone comprises five journals, which represent the most productive journals (core zone) publishing a total of 100 articles in the research area. The next step involves counting the number of journals in Zone 2: let us suppose that it contains 15 journals. Finally, imagine that the number of journals found in the third area is 45. This last group is formed by the least productive journals, as the total number of articles published by them will still be 100. Applying the basic formulation of Bradford's law to our example yields the distribution shown in Figure 4.2.

Simplifying this regularity to Bradford's formulation gives 1:3:9. In this case the constant n (more usually named k) that multiplies the number of journals across the zones will be 3.

Bradford's law can be used as a tool for collection management in libraries by identifying core journals in subject areas, thereby providing evidence for journal subscription decision-making (Wolfram, 2003). By applying this law to a given set of data it is possible to identify those journals that will account for most of the studies published in a given area.

Figure 4.2 Example of Bradford zones where $k = 3$

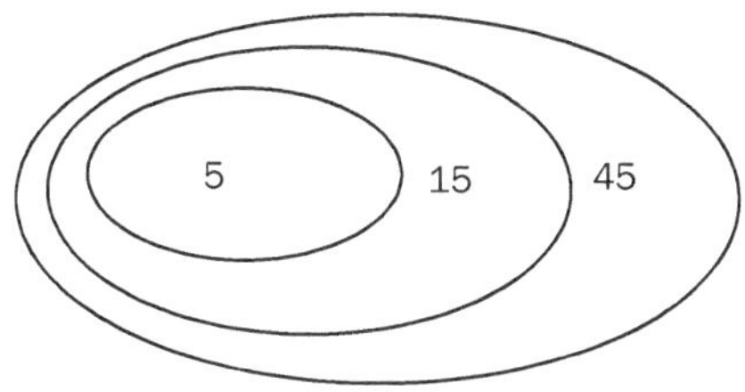

Bradford's S-shape

By considering Bradford's law in greater depth it is possible to analyse the nature of its distribution. Plotting the journal's productivity will yield a graph similar to that shown in Figure 4.3.

This figure represents the S-shape as the typical Bradford-Zipf plot (Brookes, 1969). It is the result of plotting the cumulative number of papers of each journal against the logarithm of their rank. Journals are ranked from the most to the least productive. Of course, the Bradford S-shape found in real data distributions will not be exactly like that shown in Figure 4.3, since real data will not behave in such a perfect way. However, there is evidence of data fitting the Bradford S-shape (Tsay et al., 2000).

The S-shape curve presents a phenomenon known as the *Gross droop* (Gross, 1967). This consists of a visible droop in the tail of the cumulative number of papers in the S-shape of the Bradford-Zipf plot. It is not

Figure 4.3 Bradford's S-shape

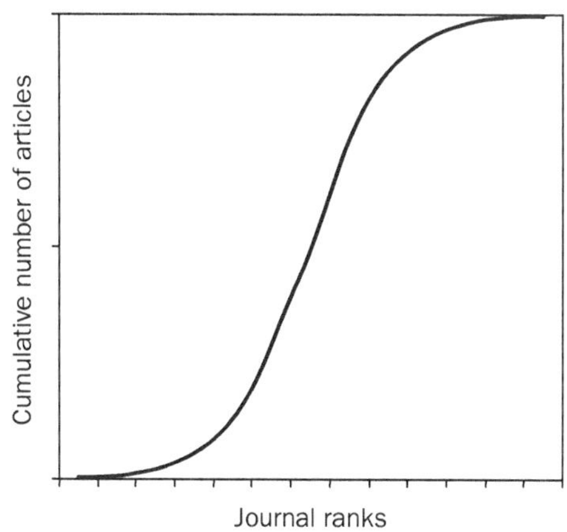

perceptible in Figure 4.3 which represents the hypothetical and ideal S-shape. The Gross droop indicates that fewer articles than expected are being contributed by the least productive journals (Wolfram, 2003). A possible explanation for this effect was proposed by Heine (1998) who argues that the observed droop is a result of an excess of low-productivity journals which contain lower than the expected number of articles.

Core journals are those that lie along the initial curved part of the S-shaped Bradford plot before it becomes a straight line.

How to apply Bradford's law

In order to illustrate how to calculate Bradford's law with real data, let us make use of the journal productivity distribution found in *automatic indexing* by Pulgarín and Gil-Leiva (2004). These authors studied scientific productivity in this area for the period 1956 to 2000, finding a total of 527 articles. As Bradford's law focuses on journal productivity it is necessary to count the number of articles published by each journal in this specific area. A frequency table showing the number of journals publishing a given number of articles can then be obtained. Table 4.1 shows the scatter of scientific literature found in the area of automatic indexing.

Firstly, from these data, it is necessary to calculate the value of Bradford's constant k, which is the multiplicator that will explain how the number of journals grows from one zone to the next. The following formula for the multiplier k of Bradford's law has been formulated by Egghe (1986, 1990a) and Egghe and Rousseau (1990):

$$k = \left(e^{\gamma} \times Y_m\right)^{1/P}$$

where γ is Euler's number ($\gamma = 0.5772$), Y_m is the maximal productivity of the journal of rank one, and P is the number of zones or Bradford groups. In the example provided, and considering Table 4.1, it can be seen that the most productive journal published a total of 66 articles, so $Y_m = 66$. In this case, P will be 4, as for this specific area it has been shown that this is the ideal number of zones for the distribution. In other words, the value of P is set at 4 because the authors state that their data fit a four-zone distribution. However, in each single case it is necessary to test the number of zones that will fit the data, according to the formulation of Bradford's law, represented in Figure 4.1. In the present example, k will be obtained as follows:

$$k = (1.781 \times 66)^{1/4} = 3.29$$

Table 4.1 Dispersion of scientific literature in *automatic indexing* area

No. of journals	No. of articles	Cumulative journals	Cumulative articles	Ln (cumulative journals)
1	66	1	66	0
1	40	2	106	0.6931
1	36	3	142	1.0986
1	28	4	170	1.3862
1	21	5	191	1.6094
1	18	6	209	1.7917
1	10	7	219	1.9459
1	8	8	227	2.0794
2	7	10	241	2.3025
4	6	14	265	2.6390
5	5	19	290	2.9444
6	4	25	314	3.2188
14	3	39	356	3.6635
27	2	66	410	4.1896
117	1	183	527	5.2094

Reprinted with permission from Pulgarín and Gil-Leiva (2004).

Secondly, it is possible to calculate the number of journals that will belong to Bradford's first group, i.e. the core journals. This is represented as r_0, and is calculated by means of the following formula:

$$r_0 = \frac{T(k-1)}{(k^P - 1)}$$

where T represents the total number of journals publishing articles in a given subject area, k is Bradford's constant and P is the number of Bradford groups. Using the data from Pulgarín and Gil-Leiva (2004), r_0 is obtained as follows:

$$r_0 = \frac{183(3.29-1)}{(3.29^4 - 1)} = 3.60$$

It can therefore be concluded that four journals will be the most productive (core journals). Knowing the values of k and r_0, we can now obtain the theoretical distribution of journals across the Bradford zones (corresponding to r_0, r_1, r_2 and r_3). The expected number of journals to be found across the zones in this example will be:

$$r_0 = r_0 \times 1 = 3.6$$
$$r_1 = r_0 \times k = 11.8$$
$$r_2 = r_0 \times k^2 = 38.9$$
$$r_3 = r_0 \times k^3 = 128.2$$

From this theoretical distribution of Bradford's law it is possible to test the exact fit of Bradford's law to our empirical data. The number of articles in each zone can then be counted, as shown in Table 4.2.

In this table the exact number of the multiplier k is also calculated, taking into consideration the real number of journals that will be included in each zone. As k values are very similar, and also similar to that calculated by the k formula (k = 3.29), it can be concluded that the data fit a four-zone Bradford distribution.

Having confirmed that the data fit Bradford's law the next step is to calculate the equation for the Bradford curve. A number of mathematical models have been developed to explain journal productivity (Brookes, 1969; Egghe, 1990b; Leimkuhler, 1967; Rousseau, 1994; Rousseau and Leimkuhler, 1987), although it is Leimkuhler's derivation that has been most commonly applied. Its formulation is as follows:

$$R(r) = a \log_e (1 + br)$$

In this formula, $R(r)$ is the cumulative number of articles published by the journals of rank 1, 2, 3 ... r (ranks represent the cumulative number of journals). The value a is a constant, calculated as: $a = y_0/\log_e k$.

Table 4.2 **Bradford zones in *automatic indexing* area**

Zones	Number of journals	Number of articles	k
Core	4	170	–
Zone 1	12	105	3.00
Zone 2	39	113	3.25
Zone 3	128	139	3.28

Reprinted with permission from Pulgarín and Gil-Leiva (2004).

y_0 is the number of articles found in each group (considering that each zone will include the same number of articles) and is calculated by means of a simple formula: $y_0 = A/P$, where A is the total number of articles found in the literature and P is the number of Bradford zones. In our example this is given as follows: $y_0 = 527/4 = 131.75 \cong 132$. Knowing this value it is then possible to obtain the a value. In this case, $a = 132/\log_e 3.29 = 110.85$.

The only value now needed to calculate Bradford's formulation is b, which is obtained as $b = (k - 1)/r_0$. As the values k and r_0 are known they can be introduced into this formula to give $b = (3.29 - 1)/3.6 = 0.636$.

Finally, as all the necessary values for Leimkuhler's formulation are known they can now be introduced into the original formula:

$$R(r) = 110.8 \log_e (1 + 0.64r)$$

This formula shows the cumulative number of articles $R(r)$ across the different ranks (from 1 to r). By replacing the different values of the cumulative number of journals r it is possible to obtain the value of the expected number of articles $R(r)$. For instance, for $r = 25$ (when the cumulative number of journals takes the value 25), we can calculate the expected cumulative number of articles $R(r)$. Then, $R(r) = 110.8 \log_e(1 + 0.64 \times 25) = 313.9$, which is very similar to the result found in the real data (Table 4.1).

Journal ranks

Another common analysis included in bibliometric research and which also studies journal productivity involves ranking journals according to the frequency of the documents they publish. Obtaining a list of journals in order of decreasing productivity is easy and provides important information. Furthermore, it is a prior step in applying Bradford's law, so it is worth performing. As stated previously, identifying those journals that publish most articles about a given subject has practical implications.

The next example concerns journal ranking in the field of *neuroscience research*. First, we have to define which documents will be included in the analysis to identify the most productive journals in this area. Here the time period was limited to the documents published in this field over the last five years, i.e. from 2004 to 2008. The bibliographic search was conducted in the ISI Web of Science and all the articles gathered fulfilled the condition of having the keyword *neuroscience* in their topic. This yielded a total of 5,862 articles published in 1,586 journals. By ranking the journals according to the number of documents published it is possible to identify the most productive ones. Table 4.3 shows the ten most productive journals in neuroscience research from 2004 to 2008.

Table 4.3 Most productive journals in *neuroscience* research from 2004 to 2008 (WoS)

Journal	Number of documents	%
Neuroscience Research	801	13.7
Science	98	1.7
Neuroimage	94	1.6
Nature	82	1.4
Lectures Notes in Computer Science	69	1.2
Nature Reviews Neuroscience	69	1.2
Journal of Cognitive Neuroscience	61	1.0
Journal of Neuroscience	61	1.0
Journal of Neuroscience Methods	61	1.0
Neuropsychologia	56	1.0

As can be seen in this table, there is one journal that clearly stands out in this research field: the journal *Neuroscience Research* accounts for almost 14 per cent of total productivity in this area.

The practical implications of identifying journal ranks have been pointed out by Lascar and Mendelsohn (2001), who analysed journal productivity in *structural biology*, a sub-discipline of molecular biology. These authors aimed to identify key library resources in their area, both printed and online, and carried out a particular data search. They identified a group of key researchers in their field and gathered the studies they had published. They then established a ranking of the most productive journals in this area. By taking into consideration those journals which publish articles by outstanding researchers in structural biology, they can then make proposals to their research centre regarding the subscription to a given online journal package.

A common feature of Lotka's and Bradford's laws

So far we have considered the main empirical laws of information science: Lotka's law about author productivity and Bradford's law regarding the scatter of journal productivity. In addition, Zipf's law was mentioned in the context of Bradford's S-shape distribution. This section reviews the common characteristics of these three laws. First, let us look at how Zipf's law is applied.

Zipf's law (Zipf, 1935, 1949) was proposed to be applied to recorded discourse, it being used in social sciences disciplines such as linguistics. Zipf stated that if one takes the words making up an extended body of text and ranks them by frequency of occurrence, then the rank of words multiplied by their frequency of occurrence will be approximately constant. This law was subsequently verified by Hanley (1937), who analysed the frequency of occurrence of the words in James Joyce's novel *Ulysses*.

Chen and Leimkuhler (1986) demonstrated that Lotka's, Bradford's and Zipf's laws have something in common, and even concluded that they were mathematically equivalent. These authors pointed out that all three laws relate, via a simple function, two variables. In the case of Lotka's law, this function explains the relationship between the number of authors and the number of papers. In the case of Bradford's law, the function considers the number of journals and the number of related papers. Finally, Zipf's law takes two other variables: the number of words in a text and their frequency of occurrence.

These authors point out that there are two main characteristics shared by these laws. The first is that all three study the particular arrangement of two groups: the observation and the class (Hubert, 1981). The variables involved will be the difference between these laws (Table 4.4).

The second common characteristic between these laws is that they all involve a relatively simple model of the particular arrangement of the observation-class groups. Lotka's law is based on a mathematical function that relates the number of authors y_x to the number of publications x; the function of Bradford's law relates the number of articles $R(r)$ to the journal rank r; and the function of Zipf's law relates the number of occurrences $g(r)$ associated with a word r.

Other authors have provided evidence for an equivalent mathematical formulation between these laws. Considering the similarities between Bradford's and Zipf's laws, Kendall (1960), Brookes (1968, 1969),

Table 4.4 Variables involved in Lotka's, Bradford's and Zipf's laws

	Observation	Class
Lotka	Papers	Author
Bradford	Papers	Journal
Zipf	Word occurrences	Word

Fairthorne (1969), Wilkinson (1973) and Egghe (1991) all offer mathematical evidence of this fact. As regards the equivalence between Lotka's and Zipf's laws, mathematical evidence has been reported by Bookstein (1976) and Egghe (2005). Finally, similarities between Zipf's and Price's laws have been demonstrated by Egghe (2005).

In summary, these laws share a similar mathematical base, being used to study different areas of productivity in bibliometric research.

5

Scientific collaborations

Scientific reports are most often the result of the work of various researchers. Indeed, and as Posner (2001) states, academic work increasingly relies on teamwork, just as industrial production does. Therefore, given that advancements in science will be the result of collective efforts, it will be interesting to study the collaboration between authors or institutions. The analysis of scientific collaborations was first introduced by Price (1963) in his book *Little Science, Big Science*, which, as we saw previously, is a seminal text in the field of scientometrics and explains the basic principles of scientific advancement. Beaver and Rosen (1978, 1979a, 1979b) can be considered the first authors to study scientific collaboration in different sub-fields. Also, as Hou et al. (2008) state, scientific collaboration can be analysed at the micro level (individuals), the meso level (institutions) or the macro level (countries). Collaboration patterns become complex, and can even be compared to the structure of neural networks or the Internet (Barabási et al., 2002). As regards the descriptive analysis of bibliometric studies we have seen that it may be interesting to count the number of authors participating in articles in order to develop a profile of the percentage of articles that have been written by one, two or more authors. This is an initial and descriptive approach about the degree of collaboration. The present chapter takes a more detailed look at collaboration in order to gain an idea of the basic tools that can be used when trying to analyse scientific collaborations in a given area of study.

Collaboration index

Let us start with a descriptive collaboration index that will provide information about the number of authors who have participated in writing articles. It is also possible to derive a trend regarding the number of authors contributing to publications in a given field over time.

In order to illustrate how to apply this descriptive index let us consider an area of study that has emerged in recent years, namely *obesity surgery*. As the prevalence of obesity, and the number of morbidly obese patients, has rapidly increased in recent years, obesity surgery has become the main treatment for weight loss in these patients. The example provided in this section is based on the scientific productivity found in the ISI Web of Science for a ten-year period (1998 to 2007) and using the keywords *obesity surgery*. Table 5.1 shows the information needed to calculate this descriptive index.

For each year it is necessary to obtain the number of articles published as well as the number of signatories found in each year. The number of signatories will not coincide with the number of authors writing the articles, since some authors participate in more than one study; furthermore, more than one author will be a signatory to the same article, as must be the case for scientific collaborations. With this information it is possible to calculate a ratio between the number of signatories and the number of articles published, and this forms part of the collaboration index, also shown in Table 5.1. This index provides an idea about the degree of collaboration in a given field of study for a particular year. However, rather than simply knowing the degree of collaboration in a given year it is more interesting to compare this collaboration index across years. In the present example, these indices take values from 3.34 in 1999 to 5.32 in 2002. This means that publications about obesity surgery for the time period selected (1998–2007) are the result of collaborative work, with an average of about three to five authors working together on the same article. A graphical representation of the trend shown by the collaboration index over time can also be obtained, as shown in Figure 5.1.

It seems that in this area the collaboration pattern has not followed a clearly increasing profile, as would be expected on the basis of Posner's

Table 5.1 Scientific productivity in *obesity surgery* field from 1998 to 2007

	Year									
	1998	1999	2000	2001	2002	2003	2004	2005	2006	2007
Number of articles	19	32	14	21	28	31	47	45	49	88
Number of signatories	75	107	55	103	149	118	201	220	210	433
Collaboration index	3.94	3.34	3.92	4.9	5.32	3.8	4.27	4.89	4.28	4.92

Figure 5.1 Collaboration index trend in *obesity surgery* field from 1998 to 2007

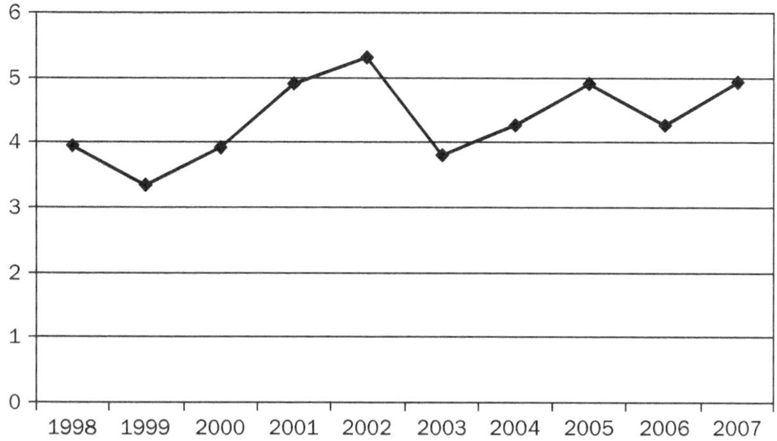

(2001) assertion that collaborative work will increase in scientific literature. At all events, this section has shown that the collaboration index is an easy way to explore the degree of collaboration or non-collaboration between authors in a research area.

National and international collaborations

Assuming that scientific literature is the result of collaborative efforts, then we can study the nature of these collaborations. While some authors will work in conjunction with colleagues from their department or university, others will work together with researchers from other institutions or countries. As such, it is possible to know the degree to which collaborations occur between people from the same institution or country.

The increasing trend toward collaboration between authors or research groups, as well as its nature, has been demonstrated by Barrios et al. (2008). These authors analysed scientific productivity in a new research field: the *psychology of tourism*. Taking into consideration the period from 1990 to 2005 they gathered a total of 572 articles from the Web of Science database about this area of study. By analysing the number of authors contributing to the articles and whether this collaboration was national or international, they were able to obtain a collaboration profile (Figure 5.2).

As is shown in Figure 5.2, the percentage of articles in which there is no collaboration has decreased over time in the psychology of tourism

Figure 5.2 **Collaboration pattern in *psychological research on tourism***

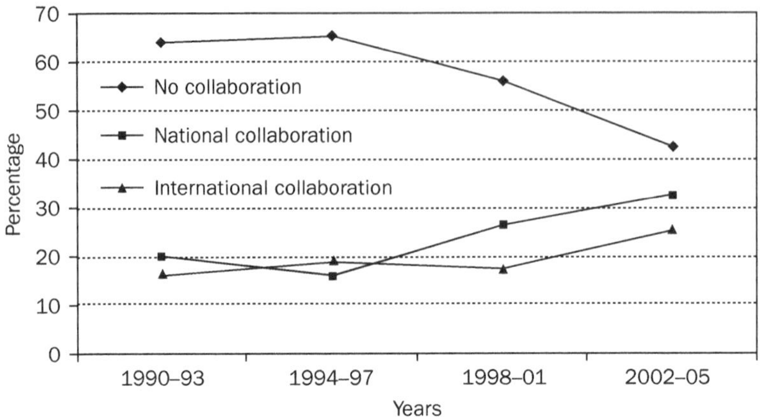

Reprinted with permission from Barrios et al. (2008).

area. In contrast, there is an increasing trend toward publications that result from collaborations between authors from different institutions, both national and international.

Further evidence of this increasing trend toward collaborative work comes from the collaboration pattern found in the area of *stem cell* research in the *neuroscience* field. In recent years there has been growing interest in the role that stem cells could play in the treatment of neurological diseases. Our analysis of the collaboration pattern in this discipline is based on a bibliographic search conducted in the ISI Web of Science using the keywords *stem cell** and *neuroscience**.[1] The year of publication was not limited. A total of 145 studies were found to have been published between 1994 and 2009, which demonstrates that this area of research is quite new. As only one paper was published in 1994, followed by a period of time in which no studies were published on this topic, this year was not included in the collaboration analysis. The year 2009 was also excluded because at the time of writing it is not possible to gather all the articles published during this year. The analysis therefore focuses on the period 1998 to 2008 and gathers a total of 141 documents. Figure 5.3 shows the collaboration pattern in this field, distinguishing over time between studies in which there is collaboration between authors and those where there is none. This pattern is calculated according to the number of documents published and, therefore, a growing publication pattern is obtained.

This figure illustrates once again the increasing interest in team work. Most of these collaborations were the result of joint work between

Figure 5.3 Collaboration pattern in *stem cell research* in *Neurosciences* (WoS)

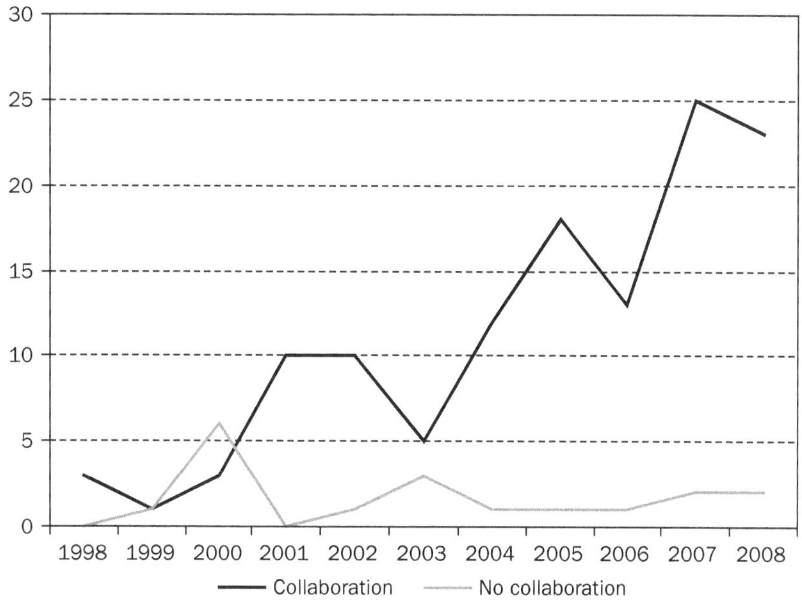

authors from the same country; only 12.2 per cent of author collaborations in this area involved people from different countries.

Both in this example about stem cell research in the neuroscience field and in the paper by Barrios et al. (2008), national collaboration is more common than that on an international scale. As Xie et al. (2008) stated, this phenomenon is bound to appear, as it is much easier to network with similar research groups attached to other institutions in the same country. These are just two examples in different areas of research about the increasing pattern of collaboration that is currently taking place.

Another way of analysing collaboration patterns is to quantify the percentage of intranational and international collaborations. An example of such an analysis is the paper by Gómez-Benito et al. (2005), which describes a bibliometric study about a specific topic in psychometrics: *differential item functioning*. The authors analysed documents on this topic that had been published over a 25-year period (1975 to 2000). First, they quantified the number of single-author articles and only included co-authored articles in the collaboration analysis. They then looked at whether collaborations were national (where the institutions of all participating authors belonged to the same country) or international (involving authors from different countries). A graphical representation of the nature of these collaborations is shown in Figure 5.4.

Figure 5.4 **Distribution of collaborations with respect to number of authors in *differential item functioning* area**

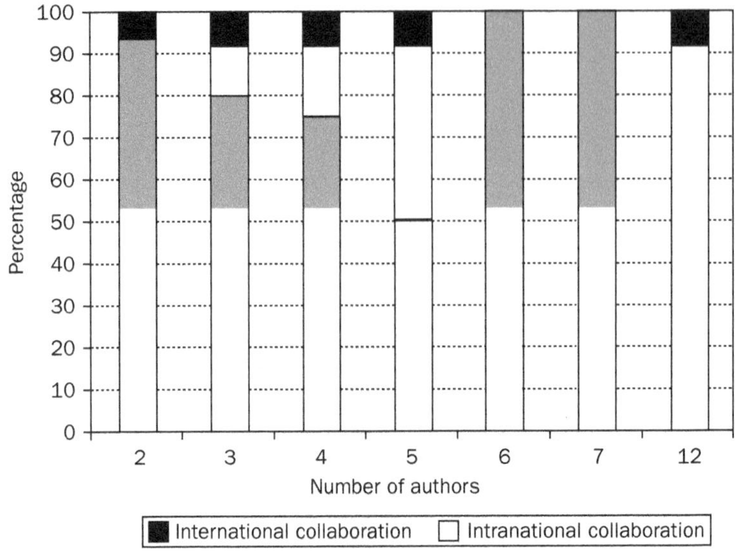

Reprinted with permission from Gómez et al. (2005).

This figure reveals a regularity for those articles in which between two and five authors have participated. Between these values the number of authors is proportional to the percentage of international collaboration. Furthermore, in those articles where more authors have participated, more international collaborations are found. However, this pattern is not found for articles written by six and seven authors, as between these values only national collaborations were found. Finally, the articles in which a total of twelve authors have participated are the result of work between authors from different countries, i.e. international collaborations.

A further step as regards the collaboration pattern is to analyse international collaborations, which would identify those countries that most often publish together.

Co-authorship index

Another possible way of analysing author collaboration patterns is the co-authorship index (CAI) (Schubert and Braun, 1986). This index is obtained by calculating proportionally the number of single-, two-, multi- and mega-authored papers for different nations or different sub-disciplines. The CAI is given by the following formula:

$$\mathrm{CAI} = \left(\left(N_{ij} / N_{io}\right) / \left(N_{oj} / N_{oo}\right)\right) \times 100$$

where N_{ij} denotes the number of papers co-authored by j authors in the i-th country, N_{io} denotes the total number of papers in the i-th country, N_{oj} denotes the number of papers co-authored by j authors in all countries and N_{oo} refers to the total number of papers in all countries.

The numerator of this formula is the proportion of articles co-authored by a given number of authors in a given country (or sub-discipline). The denominator is the proportion of articles co-authored by a given number of authors taking into consideration all the articles found (and then, all countries or sub-disciplines). As this formula is multiplied by 100 the CAI can be interpreted as shown in Table 5.2.

This theoretical explanation of the CAI has an evident practical application, as can be seen in the following example. Let us consider all the publications in a given research area, gathering the affiliations of all the authors involved in the articles. It is then possible to know which country they come from. In this case, we will use the previous example about scientific productivity in the area of *obesity surgery*. However, only those articles published on this topic in 2007 and listed in the ISI Web of Science will be considered.

Although a total of 88 articles were found according to this criterion, only 81 will be included in the co-authorship analysis since not all of them provided author affiliation information. Thus, N_{oo} = 81, which represents the total number of articles found providing the author's country. The next step is to establish the number of authors participating in or co-authoring the articles. In this case, we will study those articles that are signed by five authors, although of course the analysis could be performed with other co-authorship values. We must then count how many articles (of those found) are a result of collaboration between five authors. This is the value of N_{oj}, which in this case is 10. N_{oo} and N_{oj} are

Table 5.2 Co-authorship index interpretation

Value	Interpretation
CAI < 100	Indicates that the number of publications is lower than the average.
CAI = 100	Indicates that the number of publications corresponds to the average within a co-authorship pattern.
CAI > 100	Indicates that the number of publications is higher than the average.

both part of the denominator of the CAI formula and refer to all the articles found in the bibliographic search. The next step is to select the country that we want to study. In this example, we will consider two different countries in order to illustrate two possible values of the CAI.

Firstly, let us select publications in which an author from Germany has participated. We thus need to count how many of the articles co-authored by five authors are authored by at least one author from Germany. In the case of obesity surgery only two articles co-authored by five authors (with at least one from Germany) were found. Therefore $N_{ij} = 2$. Finally, we need to count the total number of papers found in our bibliographic search in which at least one author from Germany has participated. In this case, $N_{io} = 6$. The required values can now be introduced into the CAI formula as follows:

$$\mathrm{CAI}_{\mathrm{Germany}} = ((2/6)/(10/81)) \times 100 = 269.9$$

This CAI value of almost 270 means that the proportion of articles co-authored by five authors (with at least one from Germany) is much higher than the average for all countries. In fact, it is more than double the overall average.

Let us now consider those articles authored by researchers from the USA, once again focusing on those with five co-authors. Thus N_{oj} and N_{oo} will have the same values as in the previous example. However, we need to count how many articles have been authored by researchers from the USA (N_{io}), and also how many of these are signed by five authors (N_{ij}). In this example, $N_{ij} = 4$ and $N_{oj} = 34$. The CAI is thus given as follows:

$$\mathrm{CAI}_{\mathrm{USA}} = ((4/34)/(10/81)) \times 100 = 95.3$$

In the case of co-authorship involving the USA it can be seen that the CAI value is quite near to 100, which represents the average. It can therefore be concluded that the proportion of articles from the USA and co-authored by five authors is quite similar to the average for all countries. However, as the CAI value is a little below 100, this means that the proportion of articles written by five US researchers is slightly lower than the average.

The CAI index was used by Guan and Ma (2007) to compare the collaboration pattern in the *semiconductor literature* between five Asian countries. They identified those countries in which scientists prefer to work in large groups of five or more authors. These were the countries that obtained a high CAI regarding mega-authored papers compared to the average for all countries. They also identified another country in which scientists usually work alone, as the CAI value for single-authored papers was higher than the average found for all countries.

Research networks

In a basic approach to scientific collaborations a number of descriptive indicators can be identified regarding the number of authors participating in the articles of a research field or the mean. The percentage collaboration as a result of work by researchers belonging to the same institution or country (intra-institutional or intranational) or those who work together but belong to different institutions or countries (inter-institutional or international) can also be calculated.

These collaboration patterns give rise to a large number of relationships between authors, institutions or countries, thus establishing a research network. Bibliometrics also analyses the relationships between the factors involved in these research networks. Network analysis applied to the study of the social agents responsible for scientific publications allows us to identify the number of members in the network, the intensity of the relationship between them and the most relevant members of the network (Scott, 1991).

By way of an example, this section looks at an article that provides an in-depth analysis of research networks in *reproductive biology* (González-Alcaide et al., 2008). The method used by these authors to identify articles for inclusion in their analysis was quite particular, since their bibliographic search was based on the journals included in the first quartile of the category *Reproductive Biology* in the Journal Citation Reports and published in 2004. As this area is very prolific compared with other bibliometric studies considered in this book, the analysis of reproductive biology gathered 4,702 papers, most of them (96.75 per cent) published by more than one author. The authors were thus able to study the nature of collaborations in this area.

The first dimension on which research networks were assessed was authors. The criterion for belonging to a given network (groups of authors that are related because they have published a given number of articles in co-authorship) was to have published six or more papers in co-authorship. The main research networks of authors in this field are shown in Figure 5.5.

This figure provides a rapid overview of author relationships in this area. It can be seen that groups of authors emerged, all of whom have published at least six articles in co-authorship. A large group of 19 authors can be identified, and these form a research network. This means that they work together in the same area and combine their efforts to publish together. These groups of authors are termed clusters. In a cluster

Figure 5.5 Research networks of authors in *reproductive biology*

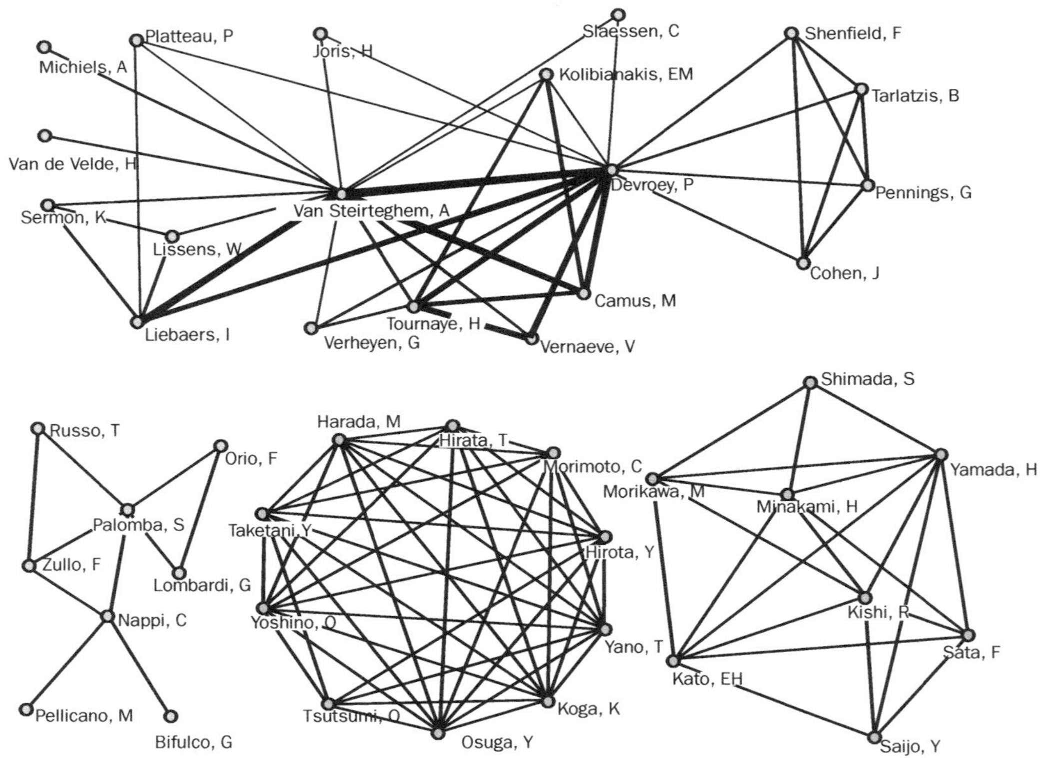

Reprinted with permission from González-Alcaide et al. (2008).

analysis, objects (in this case authors) belonging to the same cluster are more similar to one another than to objects from other clusters. We can thus identify groups of authors who are working and publishing together. The figure also shows that the central authors in a cluster will be those who establish more contacts with the other authors of the network. Those who have a stronger relationship, based on the number of articles published in co-authorship, are indicated by having a thicker line between them. In this case, the most productive authors were Van Steirteghem and Devroey and their relationship also seems to be stronger than that between other authors. Therefore they will both be located at the centre of the research network.

Another dimension studied in this paper was institutions, using the same procedure as that applied for authors. In this case, clusters or groups comprise institutions that publish together. Here the criterion to establish a collaborative relationship was that institutions had published four articles in co-authorship with others. The main research networks between institutions in this field are shown in Figure 5.6. As before, those institutions with more publications in common are shown by a thicker line joining them.

The final dimension analysed by these authors was collaboration between countries. Using the same procedure and a threshold of five or more papers written in collaboration, they obtained the following visual representation (Figure 5.7). It can be seen that the central positions of this network are occupied by the USA and the UK, which are the most productive countries. The relationship between the two is also very strong, as is that between the USA and other closely related countries in the research network.

This example has shown how scientific collaboration can be analysed from the point of view of authors, institutions or countries. A cluster analysis allows us to identify groups of researchers who are publishing together in the same area and which therefore form a research network.

Invisible colleges

The expression *invisible college* was coined by Crane (1972) to designate groups of researchers working in a similar area who maintain informal contacts with one another. They will also exchange information by means of written literature. Since it is possible to identify research networks in a given research area, invisible colleges will simply appear as part of these research networks.

Figure 5.6 Research networks of institutions in *reproductive biology*

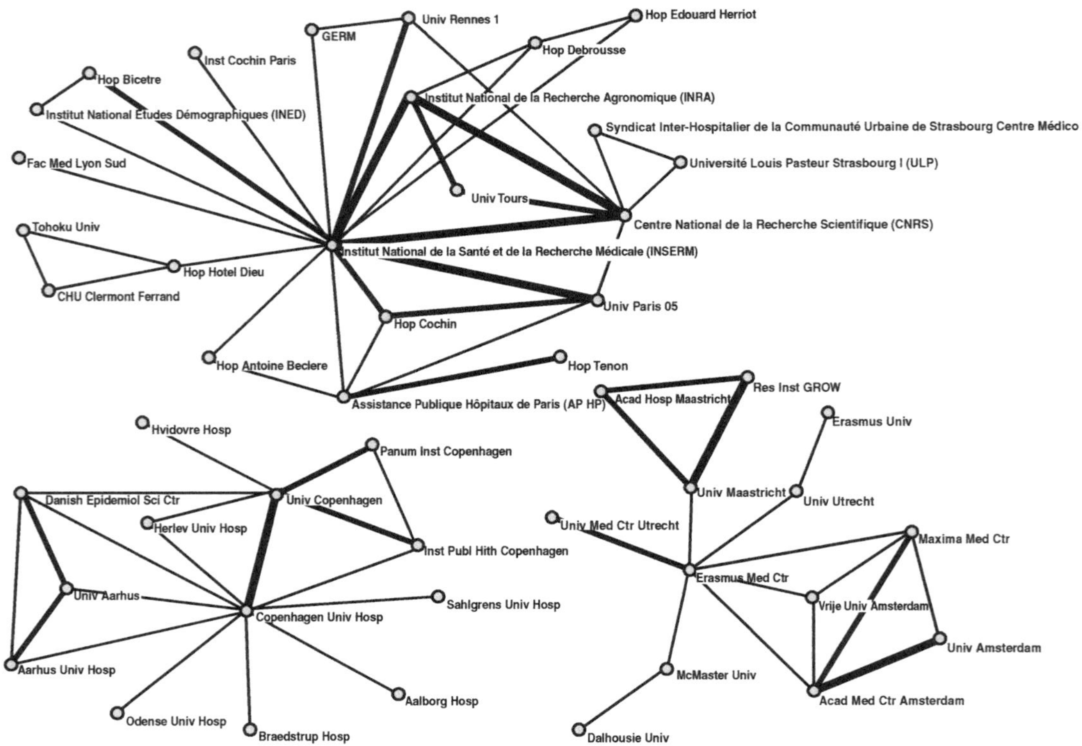

Reprinted with permission from González-Alcaide et al. (2008)

Figure 5.7 **Research network of countries in *reproductive biology***

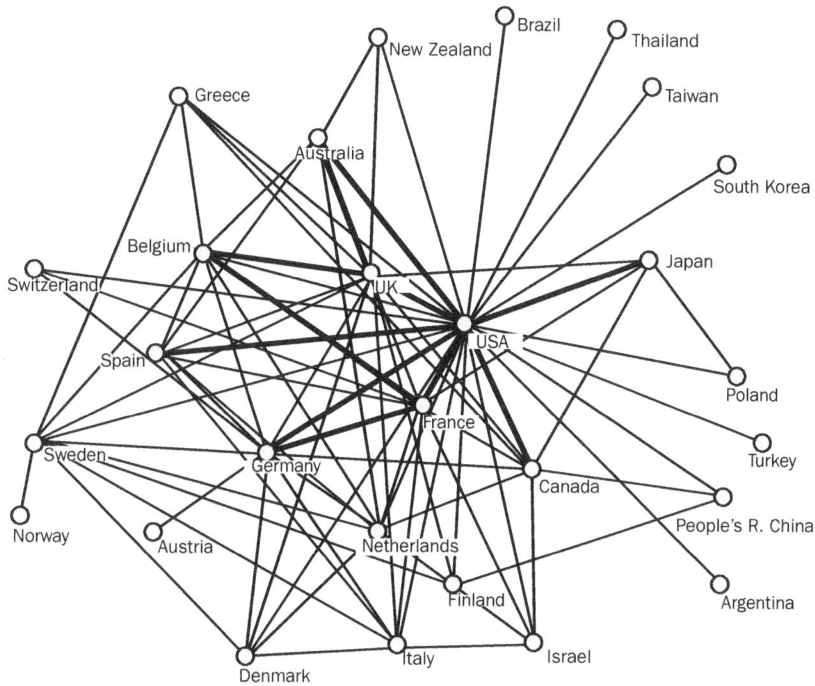

Reprinted with permission from González-Alcaide et al. (2008).

Although it seems that research networks are the result of a descriptive and static analysis of certain researchers working in co-authorship, we must go a step further to understand the influence that these groups will have. Invisible colleges are dynamic and will nowadays have an important influence on the system of scientific publication. As van Raan (2005) stated, these groups of researchers will have the ability to encourage citations between members of the same invisible college. This fact has immediate consequences, since citation analysis is the main measure used to quantify the impact – and often the quality – of a publication. The consequences of citations will be more accurately shown in the following chapters.

Note

1. By including asterisks at the ends of the keywords our search will contain entries that have the keyword or its derivatives. For example, both 'cell' and 'cells' will be included in our bibliographic search.

6

Author citation analysis

Citation analysis is a tool to identify relationships between authors or journals, as in the previous analysis of research networks. When an author publishes a study, this will include references to previous studies by other authors which are related to it. These citations reveal a connection between authors, groups of researchers, topics of study or countries. Furthermore, the impact and relevance that authors, studies or journals have on a scientific community can be measured by means of citation analysis. However, it is not advisable to use it as a single and absolute criterion for judging the importance of a publication.

Firstly, it is necessary to distinguish between a *citation* and a *reference*. Although the two terms are used interchangeably, they each represent a different view in the citing or cited perspective (Egghe and Rousseau, 1990). A reference is made within a citing document and represents an acknowledgement of another study. A citation represents the acknowledgement received by the cited document. Cited documents are generally older than the citing document, although in some cases references may be made to documents that are being published concurrently or that have yet to be published (Wolfram, 2003). Thus the fact of citing or being cited implies a relationship between two studies. However, this relationship also involves the other references given in the cited document, as these references will represent the basis of the study on which a third publication is based. Consequently, a citation analysis implies studying these complex relationships between publications which are linked by a reference or a citation.

The first paper that can be considered as a citation analysis is that published in 1927 by Gross and Gross, who studied the references found during one year in issues of the *Journal of the American Chemical Society*. However, it is Eugene Garfield who is considered to be the real pioneer of *citation indexing*. His 1955 report, in which citation indices are proposed, is considered to be an innovative paper that envisioned information tools that allow researchers to expedite their research process, evaluate the impact of their work, spot scientific trends and trace the history of modern scientific thought (Yancey, 2005). The truly innovative contribution made by Garfield was to gather studies from different sciences in the same database, thus creating a citation index in which authors could find articles from an interdisciplinary approach. Prior to this, studies of citation networks had been limited to the number of journals, the areas of scientific study and the periods of time that authors were able to consider (Garfield, 1972). These factors, together with the laborious work of compiling and manipulating manually the enormous amount of data, meant that researchers often missed important data. Also, as Yancey (2005) states, the scientific indices prior to that of Garfield were discipline-oriented, and thus researchers were not finding all the information relevant to their work. Researching a scientific area solely by its subject or keywords limits findings by ignoring relevant papers from other disciplines, as researchers in one field often publish a study that is relevant to researchers in another. This multidisciplinary citation index therefore allowed studies from different disciplines to be connected in order to aid researchers when looking for scientific papers on which to base their own work.

The first citation index introduced by Garfield (1963) and Garfield and Sher (1963) was named the *Science Citation Index* (*SCI*) and comprised a printed edition of five volumes indexing 613 journals and 1.4 million citations. As data retrieval in this format was still quite laborious it became available on magnetic tape two years later. Later, it was launched in CD-Rom format with another two citation indices: the *Social Sciences* and the *Arts & Humanities* citation indices. Finally, when these citation indices were adapted to the Web, the *Web of Science* (*WoS*) was born (Yancey, 2005). Nowadays, the *Science Citation Index* (*SCI*), the *Social Sciences Citation Index* (*SSCI*) and, of course, the *Web of Science* (*WoS*) are the main databases used in bibliographic searches, and they all belong to the *Institute for Scientific Information* (*ISI*) *Web of Knowledge* of Thomson Reuters.

With the great amount of computerised data gathered in these databases it has now become easy to access publications, references and citations in a given area of study. Although in this chapter we will focus on citation analysis as applied to authors, it will be necessary to introduce other more general concepts.

Why is studying citations of interest?

Citation measures can be as simple as a count of the number of citations received by an author, a publication or a journal during a given period of time (Wolfram, 2003), and they are an effective tool to compare research productivity and impact between authors, institutions or countries (Narin, 1976; Cronin, 1984; Borgman, 1990). At face value, citations are treated as uniformly positive recognitions of the contribution made by the author or work being cited, so the greater the number of citations an author, work or journal receives, the greater is the recognition (Wolfram, 2003). The act of citing implies a wish on the part of the author to include a given reference in his or her paper and offers significant information about performance and scientific influence (Moed, 2002a). The fact of citing a study will also affect the relationship between the cited and the citing paper, for example by arguing against the cited work's findings (Wolfram, 2003). Thus citation analysis is often used to obtain information about the impact and, more often, the quality associated with an author, a publication or a country. However, and as will be discussed later, the quality of a publication, for instance, should not only be based on the number of times it is cited, but also on other analyses.

Citation life

One of the regularities that Price (1963) found in the behaviour of science is the *obsolescence of literature*, which refers to the decline in the use of documents over time. Of course, this is related to citation analysis, since the use of a document is measured by the number of citations it receives. When a document is no longer cited it will become obsolete. Citation patterns can vary enormously depending on the documents studied.

It is possible that a large proportion of documents will never be cited, while others will receive citations in the years immediately after publication prior to becoming obsolete. It is also possible that some documents will remain uncited or rarely cited in the years following their publication, but then become recognised subsequently. This would be because, over time, more researchers become aware of the document's value and thus the number of citations will increase.

Related to the obsolescence of literature, it is also common within citation analysis to study the *age of citations*. This is assessed by knowing the year of the citation in comparison to the year of publication of the document.

An age distribution of citations is an easy analysis that shows how citations made to articles are distributed over time. The next example concerns the age distribution of citations in the field of *childhood* *epilepsy*. The time-span selected was 2003 to 2008 and the bibliographic search was conducted in the ISI Web of Science according to the following criteria: articles that included the keywords *child* epilepsy* in their title. A total of 94 articles were gathered, receiving 535 citations. Table 6.1 shows the number of publications found in each year, as well as the distribution of the citations received during this period.

As can be seen in this table, an article only receives a few citations in its year of publication as, depending on the time of year when it is published, there may not be time for it to be cited by other authors. In this example, it seems that articles receive a higher number of citations two or three years after publication. Of course, articles published before also obtain more citations. After a given period of time following publication it is expected that most articles will receive a decreasing number of citations.

Also of relevance with regard to the point at which articles receive citations is the immediacy index, as this is based on the references made in the year following an article's publication. This will be discussed further in Chapter 7.

Another perspective on citation age analysis is that taken by Chiu and Ho (2007), who studied the number of citations received by the articles they gathered about *tsunami research* from 1992 to 2004 in the Science Citation Index (SCI). These authors only included the most frequently cited articles in their area of research and showed the relationship between article life and the number of times it was cited (Figure 6.1).

The values on the *x*-axis represent the number of years since the publication date. Thus values range from 0 (representing the year 1992) to 13 (representing the year 2004). As is shown in the figure, the number

Table 6.1 Age distribution of citations in *childhood epilepsy* area from 2003 to 2008 (WoS)

Articles published		Citing year						
Number	Year	2003	2004	2005	2006	2007	2008	Total
16	2003	4	24	41	48	37	31	185
14	2004		2	11	34	26	31	104
17	2005			6	37	65	49	157
12	2006				2	22	18	42
22	2007					3	38	41
13	2008						6	6

Figure 6.1 **Citation history of the most frequently cited articles in *tsunami research***

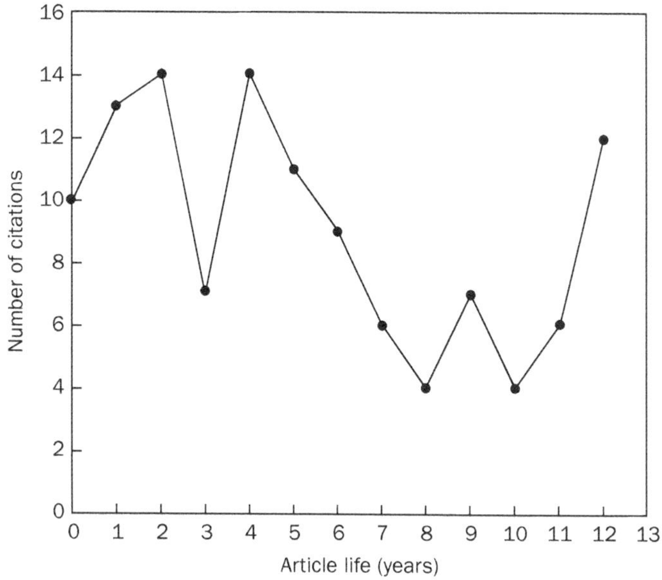

Reprinted with permission from Chiu and Ho (2007).

of times that an article is cited rises suddenly in the first year of its publication and reaches a maximum after approximately two years. After this point there is a decrease in the number of citations.

Finally, it is possible to study the lifetime distribution of citations in combination with an analysis of the type of citation. Among the citations that a study may receive, self-citations usually make an important contribution. As will be seen below, the analysis of self-citations is now becoming an important topic, in terms of both author and journal self-citations.

Glänzel et al. (2004) carried out an exhaustive bibliometric study of documents indexed in the ISI Web of Science from 1992 to 2001. Comparing the temporal evolution of self-citations and foreign citations (those made by other authors) they found that self-citations grow rapidly in the years following publication, but then decrease quite rapidly over the next few years. In contrast, foreign citations reach a peak later, but maintain a higher citation rate for longer than in the case of self-citations. Therefore it seems that self-citations age more quickly than do foreign citations.

Author self-citations

Self-citation is a common practice whereby authors cite their own previous work. A more detailed definition is that given by Snyder and Bonzi (1998) and Aksnes (2003), who state that self-citation occurs whenever the set of co-authors of the citing paper and that of the cited one share at least one author. However, data regarding self-citations must be treated with caution, as the reliability of this analysis can be affected by homonyms and thus by an erroneous count of self-citations. It can also be affected by spelling variances or misspellings of an author's name which lead to the self-citation going unrecognised (Glänzel et al., 2004).

There is no consensus about the role of author self-citations in the literature. Some authors claim that self-citations are a potential means of artificially inflating citation rates and, therefore, the author's own standing within the scientific community. However, others consider self-citations to be a natural part of scientific communication and argue that a complete lack of self-citations over a long period of time would be as pathological as an always overwhelming share (Glänzel et al., 2004).

A coherent and well-researched explanation of the role of self-citations in scientific literature is that given by van Raan (2008). His basic idea is that self-citations have a different role and function according to the two types of research groups that he proposed in a previous study (van Raan, 2006a). Here he distinguished between top-performance and lower-performance research groups when analysing the statistical properties of the bibliometric characteristics of research groups. To understand his explanation it is first necessary to clarify two terms he uses: the *size of a group* and the *cumulative advantage*. The *size of a group* is defined in terms of the number of publications that a group has. The term *cumulative advantage* is related to the *Matthew effect* proposed by Merton (1968), which has already been mentioned when discussing Lotka's law and author productivity. Applied to the present case this concept implies that larger research groups do not simply receive more citations, but do so increasingly to greater advantage.

What van Raan found was that the lower-performance groups have a size-dependent cumulative advantage for receiving citations. This means that the greater the number of publications in a group, the more those publications are promoted and the fewer the articles that go uncited. Consequently, top-performance research groups do not need internal promotion, as authors in other groups are encouraged to take notice of their publications and will cite them. Consequently, as lower-performance groups will receive fewer external citations (by authors from other research groups), a good way of promoting their work is to start with self-citations from within the group, and then to encourage external citations (Fowler and Aksnes, 2007; van Raan, 2006a). In this way, the number of non-cited publications by lower-performance groups will be reduced. It therefore seems that size reinforces an internal promotion mechanism.

The analysis of self-citation rates is becoming a topic of interest due to the influence that self-citations may have on some bibliometric indicators. The wider use of author self-citations would also influence self-citation counts within research groups, journals or disciplines.

In recent years the analysis of self-citations has been studied from different points of view. Although this section focuses on author self-citations, the topic could also be studied according to research groups or journals. As a first approximation to author self-citations it is worth taking a look at the paper by Davarpanah and Aslekia (2008). These authors analysed the author self-citation rate in the field of *library and information science* from 2000 to 2004 in the Social Sciences Citation Index. They found that almost half the total number of citations (47.6 per cent) were self-citations, and also that the number of times a work had been self-cited varied, ranging from 1 to 7 times in this case. It is clear that self-citations are a common practice among authors.

The increasing interest in self-citations is also due to the impact they have when analysing self-citations within journals. As regards the self-citation of journals, van Raan (2008) noted a similarity to group self-citations. He states that the proportion of self-citations tends to decrease with both journal impact and performance, it being smaller for those journals with a higher impact factor. These concepts will be seen in more detail in Chapter 7.

Hirsch index

The Hirsch index, more commonly known as the h-index, was proposed by Hirsch (2005) to quantify an individual's scientific output. Specifically, he tried to quantify the cumulative impact and relevance of an individual's scientific output. The index is defined as follows: a scientist has index h if h of his or her N papers have at least h citations each and the other $(N - h)$ papers have $\leq h$ citations each. As Braun et al. (2006) state, the good reception of this index in the scientific community was due, in part, to the work of Ball (2005) and Holden (2005) in the prestigious journals *Nature* and *Science*, respectively.

In Hirsch's original paper the h-index was based on authors and citations received by scientists who have won the Nobel prize. He aimed to propose a single number as a simple and useful way to characterise the scientific output of a researcher (Hirsch, 2005). The idea behind this index is that an author, during his or her research career, will have published a given number of papers which will have received a given number of citations. Thus the index takes into consideration both the quantity of papers published by an author and the number of citations these papers have received, which will reflect their evaluation by the corresponding scientific community (Zhivotovsky and Krutovsky, 2008).

Let us take a more detailed look at the meaning of the h-index and how it can be calculated. If a researcher has an h-index of 25 this means that he or she has written 25 papers, each of which has received at least 25 citations. The data required to calculate the h-index are simple and easy to obtain: the articles published by a given author during his or her career and the number of times that these articles have been cited. By ranking the papers in decreasing order according to the number of citations received we can identify the h-index as the highest rank, such that the first *h* publications each received at least *h* citations. The *Hirsch core* will correspond to all the papers ranked between rank 1 and rank *h*.

The advantage of this index is that it yields a single number which measures the broad impact of an individual's work, avoids the shortcomings of other indices and, as we will see later, allows authors to be compared according to their h-index.

The following example illustrates how the h-index can be applied to assess an author's scientific productivity. The first step is to select the author to be evaluated, and the reasons for choosing one author or another will vary. As this is only an example let us select an author who has published one of the most-cited articles in the bibliometric field. The procedure is as follows. A bibliographic search is conducted in the ISI Web of Science with the keyword *bibliometrics* in the topic, and no limits are set as regards the publication year. This yields a total of 687 documents which are then ranked according to the number of citations received, thus enabling us to identify the most-cited article. The resulting article had received 173 citations, and its first author was Stephen P. Harter. Of course, this procedure is particular to the present case.

Having selected Stephen P. Harter as the author to be studied another bibliographic search now has to be carried out to identify all the articles he has published during his career. This search is again conducted in the ISI Web of Science, this time with the keywords *Harter SP* in the author field and without limits on the publication year. As the h-index considers both his productivity and the impact of his citations, we need to gather all his published papers. The search reveals a total of 77 documents published by this author between 1971 and 2000, and these 77 papers must then be ranked according to the number of citations received. This ranking is shown in Table 6.2.

The first column shows the rank number assigned to each paper, according to the number of citations received. Naturally, the first paper is the one that received the highest number of citations, in this case 173. In contrast, the last papers in the table are those which have not been cited since they were published.

This table provides the information required to obtain the h-index. In this case, $h = 15$, calculated as follows. In this particular case the author has 14 papers with 16 or more citations each, while 15 have 15 or more citations each. However, the h-index cannot be higher than 15, as there are not 16 papers with 16 or more citations each. Specifically, the paper ranked in the sixteenth position has 14 citations, and so the highest possible value for h is 15. It can therefore be stated that this author has 15 articles with at least 15 citations each. Among these 15 articles there will be some that have a large number of citations, as in the case of the paper ranked first, with 173 citations. These 15 papers will form the Hirsch core, which is formed by the group of papers included in the h-index (from the one ranked first to that ranked in the *h*th position).

In the h-index what is important is not the highest or lowest number of citations received but, rather, the h papers obtaining at least h citations each. Table 6.2 shows how the papers ranked from the first to the fifteenth position have at least 15 citations (citations $\geq h$). In contrast, the remaining papers, ranked from the sixteenth to the last position, have fewer than 15 citations each (citations $\leq h$). This illustrates the application of the h-index (Hirsch, 2005) to our particular case. Remember that Hirsch defined it as follows: a scientist has index h if h of his or her N papers have at least h citations each and the other $(N - h)$ papers have $\leq h$ citations each.

One approach for estimating the total number of citations received by an author from his or her h-index is described by Hirsch (2005), where he estimates the total number of citations as h^2. However, and as he points out, the total number of citations will usually be much larger than the h^2 value, because h^2 will underestimate the total number of citations of the most-cited papers and ignore the papers with $< h$ citations. Returning to the example of S.P. Harter it can be seen that $h^2 = 225$, while the total number of citations obtained from the sum of the absolute frequencies of Table 6.2 is 756. Thus, as expected, the total number of citations received by this author during his career is higher than the h^2 value.

Table 6.2 Citation rank of the papers published by S.P. Harter

Paper rank	Number of citations	Year published
1	173	1992
2	64	1996
3	56	1997
4	47	1975
5	46	1998
6	43	2000
7	36	1985
8	33	1993
9	27	1975
10	26	1996
11	22	1971
12	21	1997
13	17	1984
14	16	1990
15	15	1992
16	14	1990
17	13	1979
18	13	1984
19	11	1978
20	10	1982
21	8	1998
22	8	2000
23	7	1988
24	6	1990
25	5	1978
26	5	1996
27	4	1996
28	2	1988
29	2	1990
30	2	1992
31	1	1981
32	1	1986
33	1	1993
34	1	1998
35	0	1971
36	0	1971
37	0	1975
38	0	1978
39	0	1979

Table 6.2 Citation rank of the papers published by S.P. Harter (*cont'd*)

Paper rank	Number of citations	Year published
40	0	1979
41	0	1979
42	0	1979
43	0	1980
44	0	1980
45	0	1981
46	0	1981
47	0	1981
48	0	1981
49	0	1981
50	0	1982
51	0	1982
52	0	1984
53	0	1985
54	0	1985
55	0	1985
56	0	1986
57	0	1987
58	0	1987
59	0	1987
60	0	1987
61	0	1987
62	0	1987
63	0	1988
64	0	1989
65	0	1990
66	0	1990
67	0	1991
68	0	1991
69	0	1992
70	0	1993
71	0	1995
72	0	1996
73	0	1997
74	0	1998
75	0	1998
76	0	1998
77	0	1999

Another characteristic of the h-index is that it increases over time. As Hirsch (2005) says, for a given individual one would expect *h* to increase in an approximately linear way over time, as authors publish more papers and receive more citations as their careers evolve. However, in reality this phenomenon fits an exponential function rather than a linear one. Even when an author stops publishing, the h-index can continue to increase, as he or she may continue to receive citations.

Since the h-index is based on the citation count of publications, it will be influenced by the citation life. As we saw previously, documents have a citation life, a period during which they receive citations. After this period, some of them become obsolete as they no longer receive citations. The distribution of the citations received by an article will vary depending on many variables: some articles will receive a large number of citations just after publication, while others will receive a similar number of citations but distributed over a longer period of time. If papers receive citations during a limited period of time and then become obsolete (Redner, 2005), papers contributing to an author's h-index during a given period of time will no longer contribute to the h-index later in his or her career. In fact, some papers that are included in the h-index at a given point in time may not be counted in a later analysis if other articles by the same author receive more citations (Hirsch, 2005).

In his original paper Hirsch discusses the expected *h* values that would be associated with a researcher's status or his or her membership of a prestigious scientific association. However, as has been argued throughout this book and also by Hirsch himself, several criteria should be used to assess researchers, especially in terms of knowing the implications that these evaluation procedures would have on a researcher's career.

Having looked at the meaning of the h-index and how to calculate it, it is now worth considering why it might be used. Firstly, the h-index yields a single number that gives an idea of an individual's research output, both in terms of productivity and impact. Secondly, as the index does not consider the total number of citations it cannot be inflated by a small number of big hits, which may not be representative of the researcher's career. Similarly, the presence of non-cited articles does not undermine the index. Thirdly, and related to the previous point, the h-index does not take into consideration the ratio of citations received per paper, as this procedure would penalise productivity and reward those researchers with a small number of highly cited papers. Another advantage of the index is that it takes into account all the papers published by an author, not only the most significant ones or the citations received by the most-cited ones. Finally, the h-index allows us to compare authors, as two researchers with similar h-indices are comparable in terms of their overall scientific impact.

Despite the advantages of the h-index with regard to the study of an individual's research output, there are some factors that may influence its value. In order to identify the variables that could bias the h-index, Costas and Bordons (2007) studied the relationship between this index and other bibliometric indicators. They showed that other absolute indicators of productivity, such as the total number of articles published or the total number of citations received, did influence the value of the h-index, as had been argued by previous studies (van Raan, 2006b; Saad, 2006). Obviously, the number of papers published by an author will have a direct influence on the index, as this will correspond to the maximum value that the h-index can reach. The rate of self-citations can also inflate an author's h-index (Zhivotovsky and Krutovsky, 2008). Another limitation of this index is that despite it being useful to compare the scientific output of authors, differences between fields could affect the data. These and other limitations and disadvantages of the h-index are discussed in Chapter 8, together with the limitations of other indices and aspects of citation count analysis.

So far we have looked at how the h-index can be used to assess individual productivity. However, its application has quickly spread to the evaluation of other sources of data. For example, Braun et al. (2006) argued that it was a useful supplement to assess journal impact factors. This application will be discussed in the next chapter, along with other indices applied to assess journal impact in the scientific community. Apart from author and journal assessment the h-index has many other applications, since it can be applied to other levels of productivity. In this context, it is important to mention the work of Schubert (2007), who introduced the concept of *successive h-indices*, derived by calculating an h-index from a set of other h-indices. The different levels of analysis are clearly discussed by Egghe (2008a) and are summarised in Table 6.3.

On the first level, corresponding to a micro level, the data considered for the h-index analysis will be the papers published by a given author. This yields an h-index that describes the individual's scientific output. At the second level, productivity is considered from a broader point of view, and here we would take into consideration different authors, with their corresponding h-indices, from a given institution. By ranking the h-indices of authors in decreasing order it is possible to obtain a global h-index that gives us an idea about the scientific productivity of a given institution. On the third level, more than one institution is taken into

Table 6.3 **Levels of analysis for the successive h-indices**

	Ranks	Result
Level 1	Papers	Author's individual productivity
Level 2	Authors	Institution productivity
Level 3	Institutions	Country productivity
Level 4	Countries	Global productivity

account in the analysis. By gathering the h-indices of various institutions and ranking them according to their value, a more global h-index can be obtained. If we rank an exhaustive list of the institutions of a given country, this provides an h-index of that country's productivity. Finally, on the fourth level, h-indices of countries could be ranked again so as to obtain a more global h-index about productivity in a group of countries.

The mathematical basis for modelling h-indices is provided by Egghe (2008a), who also proposes the concept of the *global h-index* of the meta-author, which comprises all the articles and their citations of the different authors in a group. In this case, papers and citations from a research group will be ranked in order to obtain a global h-index. This index will be obtained through a different procedure than that used for successive h-indices, since in this case papers will be ranked according to their citations. The global h-index will be larger than any successive h-indices of authors, institutions or countries.

g-index

The g-index was proposed by Egghe (2006a, 2006b) as an improvement of the h-index (Hirsch, 2005) for measuring the global citation performance of a set of articles. As we saw in the previous section, the h-index has some advantages when assessing the quantity and quality of papers published by an author, or even a journal. One of these advantages is that it is a robust index, since it is insensitive not only to an accidental set of uncited (or lowly cited) papers, but also to the set of highly cited papers.

However, Egghe (2006c) considered that this characteristic, far from being an advantage, is a limitation that can, nonetheless, be easily overcome. While he agrees that the insensitivity of the h-index to the lowly cited papers is an advantage, as it does not punish those authors who have a set of uncited papers, he argues that the index should be sensitive to the level of highly cited papers. Note that once a paper belongs to the Hirsch core, meaning that it has $\geq h$ citations, the number of citations it receives over the following years is not important. For instance, if an author has $h = 15$, this means that he or she has 15 articles with at least 15 citations each. Consequently, these 15 articles belong to the Hirsch core, and it does not matter whether they receive 15, 20 or 100 citations each. However, Egghe argues that the number of citations received should also be taken into consideration as a measure of the overall quality of a scientist or journal.

The g-index is based on one of the characteristics of the h-index identified by Hirsch himself (2005), who also asserted that the expected number of citations received by an author corresponds to h^2. However, as we have seen, the total number of citations is usually much higher than this value. Thus the g-index is defined by Egghe (2006a, 2006b) as the unique largest number such that the top g papers together receive g^2 or more citations. Consequently, $g \geq h$. As the g-index takes into account the highly cited papers, the more highly cited papers an author has, the higher his or her g-index will be. The main advantage of the g-index is that it clearly overcomes the problem of the h-index not including an indicator for the internal changes of the Hirsch core (Jin et al., 2007). Thus, as pointed out by Egghe (2006c), it yields a better distinction between scientists from the point of view of visibility. Finally, another advantage of this index is that the total number of documents does not limit its value, as is the case with the h-index (Costas and Bordons, 2008).

The g-index is easy to calculate, as it is based on the same data gathered for the h-index, i.e. the papers published by a given author during a particular period of time (usually his or her career), which are then ranked according to the number of citations they receive. In order to see how the g-index is applied, let us return to the example used for the h-index calculation. This example concerned S.P. Harter, the first author of the most cited article in the bibliometric field according to the ISI Web of Science. The bibliographic search of this author yielded 77 papers, the most highly cited of which obtained 173 citations. As we saw previously, this author obtains $h = 15$, which means that he has 15 articles with at least 15 citations each. However, the high citation rate achieved by this author is not totally reflected by this index. Therefore we will calculate his g-index, which considers the number of citations obtained by the most cited papers. As before, we have to rank all the articles published by this author, in decreasing order according to the number of times they have been cited. However, to calculate the g-index we need to add two more columns: the cumulative number of citations and the squared rank, as shown in Table 6.4.

Table 6.4 g-index calculation for S.P. Harter

Paper rank	Number of citations	∑ citations	Rank²
1	173	173	1
2	64	237	4
3	56	293	9
4	47	340	16
5	46	386	25
6	43	429	36
7	36	465	49
8	33	498	64
9	27	525	81
10	26	551	100
11	22	573	121
12	21	594	144
13	17	611	169
14	16	627	196
15	15	642	225
16	14	656	256
17	13	669	289
18	13	682	324
19	11	693	361
20	10	703	400
21	8	711	441
22	8	719	484
23	7	726	529
24	6	732	576
25	5	737	625
26	5	742	676
27	4	746	729
28	2	748	784
29	2	750	841
30	2	752	900
31	1	753	961
32	1	754	1,024
33	1	755	1,089
34	1	756	1,156
35	0	756	1,225
36	0	756	1,296
37	0	756	1,369
38	0	756	1,444
39	0	756	1,521

Table 6.4 g-index calculation for S.P. Harter (*cont'd*)

Paper rank	Number of citations	∑ citations	Rank²
40	0	756	1,600
41	0	756	1,681
42	0	756	1,764
43	0	756	1,849
44	0	756	1,936
45	0	756	2,025
46	0	756	2,116
47	0	756	2,209
48	0	756	2,304
49	0	756	2,401
50	0	756	2,500
51	0	756	2,601
52	0	756	2,704
53	0	756	2,809
54	0	756	2,916
55	0	756	3,025
56	0	756	3,136
57	0	756	3,249
58	0	756	3,364
59	0	756	3,481
60	0	756	3,600
61	0	756	3,721
62	0	756	3,844
63	0	756	3,969
64	0	756	4,096
65	0	756	4,225
66	0	756	4,356
67	0	756	4,489
68	0	756	4,624
69	0	756	4,761
70	0	756	4,900
71	0	756	5,041
72	0	756	5,184
73	0	756	5,329
74	0	756	5,476
75	0	756	5,625
76	0	756	5,776
77	0	756	5,929

The g-index corresponds to the highest rank for which the cumulative citation count is equal to or higher than the squared rank. By consulting the table we can see that the g-index for this author corresponds to $g = 27$. The paper ranked 27th corresponds to the last rank for which the cumulative citation count is higher than the value of the squared rank (specifically, $746 > 729$).

Comparing the h- and g-indices for this author it is clear that $g > h$. As the number of citations received by the most-cited articles is taken into consideration, one would expect this researcher to have a high g-index since he has published the most-cited paper in this area.

It follows from the above that the comparison of the h- and g-indices of a group of authors from the same area of research can lead to authors being ranked in a different order according to their individual performance. To illustrate this, let us rank the ten most productive authors from the bibliometric field according to their h- and g-indices. Having selected these scientists we can then conduct a bibliographic search in the ISI Web of Science without limit of time. Their h-indices can then be calculated and the authors ranked accordingly (Table 6.5).

Table 6.6 shows the g-indices for the same authors, and it can be seen that the rank order has now changed a little. For instance, G. Holden has been favoured by the g-index in comparison with his h-index, since he ranked fifth place according to his g-index but eighth on the basis of his h-index. In contrast, G. Lewison was ranked seventh according to his h-index, but in last place on the basis of his g-index.

Table 6.5 **h-indices for the most productive authors in *bibliometrics* field**

Author	h-index
Glänzel, W.	26
Cronin, B.	21
Moed, H.F.	21
Thelwall, M.	19
Kostoff, R.N.	17
Brahler, E.	17
Lewison, G.	13
Holden, G.	13
Janssens, F.	13
Lau, C.G.Y.	9

Table 6.6 **g-indices for the most productive authors in *bibliometrics* field**

Author	g-index
Glänzel, W.	37
Moed, H.F.	36
Cronin, B.	33
Thelwall, M.	29
Holden, G.	28
Kostoff, R.N.	24
Brahler, E.	24
Janssens, F.	22
Lau, C.G.Y.	16
Lewison, G.	15

Finally, Egghe (2006c) also proposes the measure *g/h*, which corresponds to the *relative increase of g with respect to h*. As before, this can produce some changes in order with respect to the *h* and *g* rankings (Table 6.7).

In this table, authors are ranked in decreasing order according to their *g/h* value. Higher values of *g/h* indicate larger discrepancies between the h- and g-indices. As can be seen, the author with the largest increase in the g-index with respect to his h-index is G. Holden, who was already

Table 6.7 **Relative increase of g with respect to *h* for the most productive authors in *bibliometrics* field**

Author	g/h
Holden, G.	2.15
Lau, C.G.Y.	1.78
Moed, H.F.	1.71
Janssens, F.	1.69
Cronin, B.	1.57
Thelwall, M.	1.53
Glänzel, W.	1.42
Brahler, E.	1.41
Kostoff, R.N.	1.41
Lewison, G.	1.15

the most favoured by the g-index calculation. In contrast, G. Lewison has obtained the smallest increase of his g-index with respect to the h-index and is thus the least favoured by the g-index calculation.

Costas and Bordons (2008) applied this analysis to their data and found that big producers are favoured by the h-index, while scientists with intermediate levels of production obtain better results by means of the g-index.

The amount of literature reviewing both the h- and g-indices is evidence of the clear interest in assessing individual productivity. Specifically, the g-index has been studied from the point of view of its mathematical basis (Egghe, 2007, 2008b; Guns and Rousseau, 2009; Woeginger, 2008), as well as its relationship with other indices designed to improve upon the h-index (Burrell, 2009; Rousseau, 2008).

Some authors have proposed other possible applications of the g-index. For example, Tol (2008) developed the successive g-index, which is similar to the successive h-index (Schubert, 2007) discussed previously. Also, other authors have studied the application of this index to specific situations. Noteworthy in this regard is the work of Costas and Bordons (2008), who tested whether the g-index is more sensitive than the h-index when assessing selective scientists. In a previous study (Costas and Bordons, 2007) these authors found that the h-index was not able to identify those researchers who are very selective when choosing journals, and who have intermediate levels of production but with a high international impact. Their research shows that the g-index seems to be better able to assess selective scientists, since they found significant differences in the positions occupied by these scientists in the h- and g-index ranks, who also had a higher *g/h* ratio. However, they also found that productive authors are favoured by the h-index rather than the g-index.

The role of the number of citations and its relationship to the g-index has also been a topic of study. In this regard, it is worth noting the excessive influence that highly cited papers can have on the g-index, as occasional big hits are unlikely to represent a scientist's average performance (Costas and Bordons, 2008).

Related to this factor is the number of self-citations received by a paper. In this regard, Schreiber (2008) warns of the influence that self-citations have on the g-index calculation. Since this index takes into consideration the number of citations received by the most cited papers, self-citations will have an even greater influence here than in the h-index calculation.

Finally, it is important to emphasise that the g-index should be complemented by other indicators when assessing an individual's performance.

Co-citation analysis

The previous analysis of research networks showed how cluster analysis can be used to identify groups of authors with publications in common. In this section we will see how a similar procedure can be used to analyse co-citations.

Author co-citation analysis was first proposed by White and Griffith (1981) and implies that two authors are cited in the same document. Co-citation analysis has since become a common method for identifying authors in related fields (Egghe and Rousseau, 1990), as it can be assumed that there is a relationship between two references cited in the same article. If two references have been considered worthy of inclusion in the same study, one would expect them to have something in common not only with the study that includes them but also with one another. In summary, this analysis is based on the idea that co-citations represent an indicator of proximity in terms of content (Gmür, 2003). Furthermore, the fact that two authors are not co-cited frequently would seem to indicate intellectual distance (Kreuzman, 2001).

Author co-citation analysis can be useful for finding related authors when one author's name is known, thus enabling us to identify, examine and trace the intellectual structure of an academic community (Su et al., 2009).

As Gmür (2003) states, a prior step to conducting a co-citation analysis is to pre-select those authors that will be included in the analysis; this can be based on the literature of the given field or by consulting experts able to identify key authors who should be included in an analysis. This selection should subsequently be checked against the citation count found in the corresponding database.

The procedure for a co-citation analysis can be based on different statistical methods: cluster analysis, multi-dimensional scaling or factor analysis (Liu, 2005). All of these methods will use the frequency count of data pairs (co-cited authors) in order to identify their relatedness.

Traditionally, author co-citation analysis has taken into account the first author of each citation, without considering the contributions of other co-authors. As Su et al. (2009) point out, the traditional approach may overlook authors who tend to place themselves second or later in the author lists. They estimate that the percentage of authors who are never listed as the first author in the citation bank they analysed may exceed 60 per cent. Thus a significant amount of information will be missing when basing the co-citation analysis on the first author of citations. In this regard it has been argued that all the authors contributing to a study should be included in a co-citation analysis (Persson, 2001) and this would clearly offer a different and broader point of view.

The most widely used method to study co-citations has been based on cluster analysis, which yields the most commonly co-cited clusters or groups of authors. However, this procedure does not consider the possibility of an author having multiple expertise. In cluster analysis, each element (authors in this case) belongs to a given cluster and, as such, the same author cannot belong to two or more different clusters. This problem can be solved with other methods such as factor analysis, which does allow the same author to belong to different factors. Su et al. (2009) also proposed an algorithm to identify groups of co-cited authors who have published across different speciality domains.

The practical application of author co-citation analysis is evident in the work of Kreuzman (2001), who aimed to use this quantitative method to illustrate the division between the *Philosophy of science* and *Epistemology*. The documents included in this bibliometric study were obtained from the Arts & Humanities Citation Index of the ISI Web of Knowledge and covered the period 1980 to 1993. Only some of the authors in this area (a total of 62) were included in the analysis, as their work was representative of the area of study. Co-citations were analysed for all the possible combinations of author pairs in which the order of authors did not matter. This is because the number of times that authors *A* and *B* are co-cited is the same as the number of times that *B* and *A* are co-cited. Five groups of authors can be identified as having more co-citations in common and it can therefore be assumed that their work is related. This analysis also enabled Kreuzman to identify two main groups of authors according to the content of their work: one focused on philosophy and the other on epistemology.

Cluster analysis applied to author co-citations enabled this author to obtain empirical evidence about the division between two disciplines, a fact that had been previously identified by means of qualitative analysis.

Citations and research groups

The analysis of author citation leads on to the study of the citations received between groups of authors. Indeed, if we are studying the number of citations received by an author, it is also possible to analyse which authors are making them. Since citation analysis can be applied to different levels of productivity (Hou et al., 2008), in the present section we are focusing on the analysis of citations with the aim of identifying research groups of scientists. The reason for restricting the study to collaborations between authors in order to identify scientific collaborations is because other levels do not adequately reflect the trends in cooperation between individuals (Kretschmer, 2004).

Citation analysis can also be used to identify invisible colleges (Casey and McMillan, 2008). As mentioned previously, invisible colleges are networks of researchers working in a similar area and who maintain informal contacts. Authors from an invisible college will refer to each other in their documents without being linked by formal organisational relationships (Crane, 1972).

Network analysis is a good analytical tool to identify research networks based on the number of citations received. This procedure has also been discussed in a previous section (see the section on research networks in Chapter 5) when identifying groups of authors that have published in collaboration. In the present case the analysis is similar, except that the data are based on the number of citations received. As Wasserman and Faust (1994) emphasise, research network analysis provides a diagrammatic representation of the relative distance between nodes, illustrating the structural patterns and positions within the network. This procedure also reveals the number of interactions between authors, and the closeness of the relationships between nodes in a network (Casey and McMillan, 2008). From a citation map based on co-citations we can identify those authors who are closer to the centrality, and who therefore are better placed to have independent access to other members of the network (Freeman, 1979). By being close to all other authors a given researcher can disseminate information quickly through the network (Casey and McMillan, 2008). Another measure that can be obtained from network analysis is *betweenness*. Scientists with a high betweenness centrality measure are those who facilitate exchanges between less central authors, and who also have the ability to control other scientists (Scott, 1991).

At all events, it should be borne in mind that although we are focusing on network analysis applied to authors in order to identify research groups, other variables, such as journals or articles, could also be taken into account.

The fact of publishing in collaboration with other authors has direct implications for citation analysis. Narin et al. (1991) studied scientific collaborations between European countries and found that those papers which were the result of collaboration between scientists from different countries were significantly more cited than were papers of scientists from the same institution within a single country. In summary, and as Inzelt et al. (2009) state, there is evidence that scientific collaborations, particularly papers written in co-authorship, are increasing the citation rate of publications. These authors developed an index, known as the *incremental citation impact*, to quantify the citation success of the cooperation between two institutions. This index is related to an idea pointed out by Glänzel and Schubert (2001), who introduced the concept of hot links and cool links between countries. Hot links denote success, in terms of citation impact, between a pair of countries. In contrast, a cool link denotes an unsuccessful collaboration between a pair of countries.

The *incremental citation impact* developed by Inzelt et al. (2009) takes into consideration a pair of institutions. The analysis relates the number of citations and papers published by a given country to the citations received and the articles published in collaboration with another institution. The analysis is carried out by taking into consideration a given institution in relationship to another institution denoted as *k*. Its formula is as follows:

$$\Delta_k = \frac{z_k - \sum_i n_{ki}(z_i/n_i)}{n_k}$$

In this formula, z_k is the number of citations received by the articles published by a given institution in collaboration with another institution *k*, while n_{ki} is the number of articles published by these two institutions in collaboration. The value of z_i corresponds to all the citations received by the institution in question during the period selected, whereas n_i is the number of papers published by the institution in question.

The incremental citation impact can easily be obtained by counting the number of articles published jointly or separately by the two institutions, as well as the citations received by their papers. The sub-index *i* denotes the fact that a period of time can be included in the study, not just a single year. Thus counts for each of the years selected can be carried out and included in the formula under the summation Σ.

When interpreting this index a positive Δ_k indicates a citation success in the collaboration, or a hot link as Glänzel and Schubert (2001) stated. A value of $\Delta_k = 0$ indicates that institution *k* has no appreciable effect on the citation impact of the main institution studied. Finally, a negative value of Δ_k implies an unsuccessful collaboration between these institutions, as institution *k* fails to promote the citations of the main institution under study.

Self-citations among research groups have also been studied. As we have seen, author self-citation is a common practice when publishing, and this behaviour has a direct impact when studying groups of scientists working in collaboration. As van Raan pointed out (2006a, 2008), top-performance groups usually have not only more citations but also more self-citations than do low-performance groups. Thus it seems that size enhances self-citations more than it does external citations.

In relation to self-citations in research groups, Della Sala and Brooks (2008) demonstrated that a variable influencing the number of self-citations received is the *number of authors* contributing to a paper. Given the tendency for authors to self-cite their own articles, one would expect multi-authored papers to receive more self-citations. Consequently, the average number of authors per article in a subject area will correlate with the average impact factor of that area (Amin and Mabe, 2007). These authors wondered whether the number of authors signing an article is related to the total number of citations that the article receives. They put this hypothesis to the test using the 2007 ISI impact factor for the journal *Cortex*. Firstly, they identified all the substantive articles that were involved in the impact factor calculation of that year. Then they divided the articles into three groups depending on the number of authors contributing to the articles: few authors (one or two), three authors or multi-authored papers (four or more). The results showed that multi-authored papers received a significantly higher number of citations compared with few-author papers. Furthermore, and as predicted, the multi-authored papers received more self-citations than did the few-author papers. It therefore seems that writing a paper in collaboration with other authors implies a greater likelihood of being cited in the future.

7

Journal citation analysis

Citation analysis can be applied to different units of study. The previous chapter looked at citations among authors, where citation count is used as a criterion for individual evaluation. However, other basic indicators such as productivity can also be used to assess journals (Tsay, 2009). One such measure would be the number of papers published by a journal in a specific field over a particular period of time, although another common measure to assess journals is based on the citations they make or receive.

Indeed, the citation analysis of a journal usually has practical consequences. A journal containing articles that are often cited will have a wider spread, as authors have decided to include a specific reference to it in their study. As a result, readers from the same research field will be aware of the existence of this particular journal, which is thus more likely to be cited again. Consequently, the citation frequency of a journal reflects its value and the use made of it (Garfield, 1972). However, it does not mean that only the most useful journals will be the most often cited, as others may not be cited frequently despite their value in a research field. Thus care must be taken when drawing conclusions on the basis of citation analysis. The citation of journals will be influenced by their specific characteristics and, as Garfield states (1972), if we consider that all the articles published have the same likelihood of being cited, it should follow that the more articles a journal publishes, the more frequently the journal will be cited. Therefore the citation frequency of a journal will not only be a function of the scientific significance of the material published, but will also depend on the number of articles it publishes. As such, journal productivity is a factor that influences citation count. In this regard, Magyar (1974) related citations with Bradford's law. This author found that the most cited articles belonged to the most productive journals, those situated at the top of the Bradford core zone. There is also evidence to support a correlation between the number of articles published by a journal (productivity) and the number of citations (Boyce and Pollens, 1982). However, despite the evidence that journal productivity can influence citation analysis, the latter is being widely use to assess journals without taking this productivity into consideration.

The present chapter considers the main indices used to analyse journal citation, and also discusses their practical implications. In fact, journal citations have come to represent one of the main factors involved in scientific practice. Most of the indices presented in this section are based on data obtained from the ISI Web of Knowledge. As this indexation system is widely applied at present, it is important to be clear about the meaning of these indices and how can they be obtained. Data regarding the citation of journals is available in the Journal Citation Reports (JCR) from the ISI Web of Knowledge, this being based on its database of journals and articles. Of course, a lot of journals are not included in this database as they have less international impact on the whole scientific community. However, as stated above, this does not necessarily mean that they are less useful. Rather, those articles that are not included in the ISI Web of Knowledge are simply cited less often and would have less impact on the global scientific community, even though they may be really useful in a given research area. Furthermore, and returning to the relationship between journal productivity and the citation count, there is no evidence to suggest that articles published in the less productive journals of a research field are of lower value (Tsay, 2009).

Immediacy index

The *immediacy index* is an indication of the speed with which items published in journals are incorporated into other literature's references (McVeigh, 2004). This index is calculated for journals, each of which will have its own immediacy index. In a previous example (Table 6.1) we saw how article citations were distributed over time, it being shown that some articles are cited during the same year of their publication. In this regard, the more citations a journal's papers receive in the year of their publication, the higher its immediacy index will be. A high immediacy index indicates that the content of this journal is quickly noticed, highly valued and topical within the field of study (Davarpanah and Aslekia, 2008).

The immediacy index is calculated by the following formula:

$$\text{Journal immediacy index} = \frac{\text{Number of citations given to articles published in a given year}}{\text{Number of articles published in that year}}$$

However, this index can be biased by various factors. For instance, an article published at the beginning of the year is more likely to be cited during that year than is another article published at the end of the same year. Obviously, this will affect the immediacy index of the journal where the latter is published. The same applies to journals that publish infrequently in the year, as they will probably receive fewer references during the same year in which their articles are published. Another factor to consider is the size of the journal in terms of the number of articles published during a year. Small journals have an advantage in this regard, as the denominator of the formula (total number of articles) will be smaller, which may bias the index. However, it is also the case that large journals (which are disadvantaged by the denominator of the formula) publish a great number of articles and, therefore, stand more likelihood of being cited.

Let us consider two examples in which the immediacy index is applied. The first involves the *New England Journal of Medicine* and data provided for it in the 2007 Science Citation Index (SCI) of the ISI Web of Knowledge, this being the latest available version of this indexation system. During 2007 this journal published a total of 343 articles which received a large number of citations during this same year. Since a total of 4,103 citations were counted for the year 2007 the immediacy index for this journal is:

$$\text{Immediacy index}_{\textit{New Engl J Med}} = 4{,}103/343 = 11.962$$

This shows that the articles in this journal seem to receive a high number of citations in their year of publication, followed by a rapid dissemination of the studies published, with an average of almost 12 citations per article. This will therefore be considered a large journal which publishes a large number of articles that also receive a lot of citations.

The second example concerns the journal *Annual Review of Immunology*, and once again the data regarding its productivity are taken from the 2007 Science Citation Index. This is a small journal, since only 27 articles were published during 2007. However, this small group of articles has received a great number of citations during this same year: a total of 300. The immediacy index for this journal is therefore:

$$\text{Immediacy index}_{\textit{Ann Rev Immunol}} = 300/27 = 11.111$$

This journal also has a rapid influence within the scientific community, as its articles are frequently cited shortly after their publication.

These examples concern two journals that have a similar immediacy index and thus a similar average number of citations received during the year in which articles are published. This is despite the fact that the first journal is large, with many more articles published compared to the second, smaller one.

Although this index is widely used to assess the immediate impact of a journal, it is limited by the fact that it can be influenced by other variables such as the scientific field to which the journal belongs, the journal's self-citation rate or the language in which articles are written. These limitations are discussed further in Chapter 8.

Journal impact factor

The *journal impact factor* was developed with the idea of identifying the most relevant journals in an area of study. The first authors who sought to measure what is nowadays called the impact factor were Gross and Gross (1927). They were interested in identifying which scientific periodicals were needed in a college library in order to prepare students successfully for advanced work and soon realised the importance of compiling a list of relevant journals. However, it is essential to gather these journals in an accurate way. Using a subjective approach, whereby the compiler decides which journals are indispensable and the most relevant in a given discipline, would lead to an inaccurate selection, since the compiler may be influenced by needs, likes and dislikes. As a result, these authors hit on the idea of calculating a journal impact factor. As Archambault and Larivière (2009) state, the aim of this first impact factor was to facilitate the task of journal selection using objective quantitative methods. Initially, impact factor methodology was limited to a single field and references were often compiled from a single key journal or key reference monograph. However, after some years the method came to include several journals from different source fields, as well as many non-US source journals (Archambault and Larivière, 2009).

Nevertheless, it was not until the publication of two key articles (Garfield, 1955; Garfield and Sher, 1963) that the relevance of the impact factor began to be more widely recognised. Eugene Garfield is considered to be the father of the impact factor being used at present, as it was he who applied this methodology to the current indexation system. In 1963, the Science Citation Index was developed by Garfield and Sher (1963), gathering the core group of large and highly cited journals. Taking into consideration only the number of citations received by a journal, small journals (in terms of articles published) would not be selected (Brodman, 1944). However, sorting journals by impact factor enables the inclusion of many small but influential journals. Nowadays, the journal impact factor is annually calculated in the Journal Citation Reports (JCR), with the journals indexed in the ISI Web of Knowledge (Thomson Scientific). The impact factor has been applied to assess the impact of authors, groups of authors such as academic departments and disciplines. However, it is most commonly applied to journals.

The journal impact factor is obtained by dividing the number of citations received by the journal in the current year to items published in the previous two years by the number of documents published during the previous two years. Its formula is thus as follows:

$$\text{Impact factor} = \frac{\text{Citations to recent items}}{\text{Number of recent items}}$$

The impact factor will therefore represent the average number of citations received in a given year by the articles published in a journal.

However, it is important to be clear about which citations are included in the numerator of the journal's impact factor and the item count in the denominator of this formula. In the first formulation of the impact factor (Gross and Gross, 1927), self-citations of journals were excluded from the analysis in order to avoid the overestimation of citations to items published by the journal itself. However, at present the impact factor calculated by the Journal Citation Reports considers both citations and self-citations. Thus the number of citations corresponding to the numerator of the formula gathers all citations and self-citations that the journal receives. However, the denominator of the impact factor formula is the result of a quite different count, because not all the documents published in a journal are taken into account when calculating the impact factor. Since only substantive articles published in journals are considered in the Journal Citation Reports count, only original research articles and review articles are taken into account. However, other documents such as correspondence, letters, commentaries, perspectives, news stories, obituaries, editorials, interviews and tributes are not included in the count. This different way of counting items in the formula's numerator and denominator has been the object of considerable criticism, as will be seen in the following chapter.

Let us now consider two examples of the impact factor calculation for, respectively, the journals that have received the highest impact factor in the Science Citation Index and the Social Sciences Citation Index during 2007, the latest year for which data are available. Of course, both databases are included in the Journal Citation Reports but they consider different areas of research.

The journal that has received the most citations per item in the 2007 Science Citation Index (SCI) is *CA: A Cancer Journal for Clinicians*. As we are taking the year 2007 as our reference and we want to know the impact factor of this journal for this year, we must first gather the data needed to calculate it. Table 7.1 shows the number of citations received by this journal during 2007, as well as the number of items published.

As can be seen in this table, the number of substantive articles published by this journal during 2005 and 2006 is quite small (only 39). However, these articles have had a huge impact as they have received a lot of citations. Given that we wish to assess the impact that this journal had in 2007 we need to count the references made in 2007 to articles published in the two previous years (2005 and 2006). The impact factor will thus be calculated as follows:

$$\text{Impact factor}_{\mathit{CA\ Cancer\ J\ Clin}} = 2{,}692/39 = 69.026$$

This index indicates that the articles published in this journal during the previous two years have received an average of 69 citations per article during 2007. However, it should be remembered that this is the journal that has had the greatest impact in 2007 according to the Science Citation Index and the impact factor of the vast majority of journals will be lower.

The second example concerns the journal that has obtained the highest impact factor in the 2007 Social Sciences Citation Index (SSCI), in this case *Behavioral and Brain Sciences*. As before, to calculate its impact factor we need to know the number of citations received to articles published during 2005 and 2006, as well as the number of items published in this period. Table 7.2 shows this information.

Using this information the impact factor for this journal is given by:

$$\text{Impact factor}_{\mathit{Behav\ Brain\ Sci}} = 454/26 = 17.462$$

In this case, articles published during the previous two years have received an average of around 17 citations each. This impact factor corresponds to

Table 7.1 **Citations received and items published in *CA: A Cancer Journal for Clinicians* (2007 JCR)**

	2006	2005	Total
Citations in 2007 to items published in	1,361	1,331	2,692
Number of items published in	19	20	39

Table 7.2 **Citations received and items published in *Behavioral and Brain Sciences* (2007 JCR)**

	2006	2005	Total
Citations in 2007 to items published in	98	356	454
Number of items published in	12	14	26

the journal with the highest impact, as it would have received more citations per item than the other journals listed in the Social Sciences Citation Index.

Now let us compare the impact factor obtained by the first of these journals, *CA: A Cancer Journal for Clinicians*, corresponding to 69.026, and that found for *Behavioral and Brain Sciences*, 17.462. At first sight it is not clear why there should be such an enormous difference between the two, as they are both considered to be the journals with the greatest impact in their respective indices in the Journal Citation Reports. However, the reason for this discrepancy is simple: it is not possible to compare the impact factor of a journal between different areas of research because this can vary tremendously between fields. Indeed, there is evidence of high impact factors obtained by those journals whose topics are helping researchers in the biomedical field, as these fields have high citation rates and, therefore, high impact factors. In contrast, even the best and most highly recognised journals from other fields such as social sciences or mathematics obtain only small impact factors because of the lower citation propensity in these fields (Archambault and Larivière, 2009). It thus seems that impact factor disadvantages certain disciplines, because bias is implicit in the citation sample on which the journal impact factor is based (Vanclay, 2009).

However, the impact factor is more than a single number describing the impact of a journal. In practice, we can identify publications that have a strong impact during a given period of time. An example of such an approach can be found in the case of *tsunami research* (Chiu and Ho, 2007). These authors analysed the impact factor of journals publishing articles about tsunami research from 2001 to 2005, during which time there was an event that clearly influenced the level of interest in the subject: the devastating tsunami that hit Indonesia in December 2004. By analysing the distribution of articles about this topic during the time period studied, these authors identified a pattern in the impact factor of publications. During 2005, after the Indonesian tsunami, more documents were published in higher impact factor journals of this field. This is a clear example of how publications in a research field can have more or less impact over time, and also shows how public interest can promote the publication of articles in journals with higher impact factors.

At present, the impact factor of a journal has important practical consequences. In addition to helping libraries decide which journals to purchase, journal impact factors are also used by authors to decide where to submit their articles. As a general rule, the journals with high impact factors are considered to be the most prestigious (Garfield, 2006). However, and as we have seen, the impact factor has to be interpreted carefully, since factors such as the research field may affect the impact factor value. Nevertheless, journal impact factor is the most widely used and perhaps the most controversial citation measure applied to journals (Wolfram, 2003). In fact, a great amount of literature has criticised the bias, limitations and contradictions of impact factor calculation and its use. The limitations of the impact factor and their consequences will be discussed further in Chapter 8.

Let us now look at one of the criticisms levelled against the journal impact factor and then consider another index that is used in its assessment. The criticism is based on what is referred to as the *target window* of the impact factor. This corresponds to the period of time selected for which citations and items will be taken into account to calculate the impact factor. In the impact factor formula shown above, the target window was set to the previous two years. However, several authors have argued that a target window of two years distorts the index (Moed et al., 1998), even when comparing journals belonging to the same field (Glänzel and Schoepflin, 1995). As Moed et al. (1998, 1999) state, the age at which impact reaches its highest value is not necessarily during the two years following publication. Indeed, these authors consider that the maximum average impact of journal documents is only reached after two years for a limited number of journals. For these journals the impact factor calculated by the ISI Web of Knowledge covers the best years in terms of impact, but there will be other journals that obtain their highest impact three, four, five or even more years after the publication date. These will be journals that accrue citations slowly over a long period of time. It therefore seems that the impact factor is distorted in favour of journals that accumulate many citations in a brief period of time after the publication of its articles. However, the distribution of citations can show different patterns when comparing journals or disciplines. Consequently, there is a strong support for reformulating the impact factor (Monastersky, 2005) or resetting the target window at three or more years (Glänzel and Schoepflin, 1995; Moed et al., 1999). In fact, this suggestion was already made back in 1988 by Rousseau, who considered that the impact factor should be based on a five-year observation period rather than the two-year period currently applied.

Despite this strong criticism, Garfield (1999, 2006) considers that a two-year target window provides a measure of a journal's current impact, and that this current aspect of the measure would be lost by selecting a longer time period. In fact, he argues that when studying journals within disciplinary categories, the rankings based on one-, seven- or 15-year impact factors do not differ significantly (Garfield, 1998a, 1998b). However, the arguments against a two-year impact factor have been taken on board to some extent and, since February 2009, a *five-year journal impact factor* is now included in the Journal Citation Reports from the ISI Web of Knowledge. This supports the idea that citation behaviour varies across disciplines, and while some fields advance very quickly, research in other fields makes its impact more slowly over an extended period of time. Life sciences are a good example of fields in which published research gets cited relatively quickly, while others as mathematics will be cited more slowly after the publication of articles (Thomson Reuters, 2009).

The five-year impact factor formula is similar to that of the traditional impact factor, but it is based on citations in the current year to journal material published in the previous five years. It is calculated as follows:

$$\text{Five-year IF} = \frac{\text{Number of citations in the current year to items published in the previous five years}}{\text{Number of articles published in the same five years}}$$

The five-year impact factor provides a measure of journal impact in fields where the influence of published research evolves over a longer period of time than would be captured by the traditional two-year index (Thomson Reuters, 2009).

Let us return to our previous examples and calculate the five-year impact factor for the *CA: A Cancer Journal for Clinicians* and *Behavioral and Brain Sciences*. This time, we need to gather the number of citations received in 2007 to items published during the previous five years, as well as the number of citable items.

This information with respect to the *CA: A Cancer Journal for Clinicians* is shown in Table 7.3.

The five-year impact factor is calculated by obtaining the total number of citations received in 2007 to items published from 2002 to 2006 and dividing it by the total number of items published during this period:

$$\text{Five-year IF}_{\textit{CA Cancer J Clin}} = 4{,}105/90 = 45.611$$

Table 7.3 **Citations received and items published in *CA: A Cancer Journal for Clinicians* from 2002 to 2006 (2007 JCR)**

	2006	2005	2004	2003	2002	Total
Citations in 2007 to items published in	1,361	1,331	496	347	570	4,105
Number of items published in	19	20	18	16	17	90

Let us now apply the same procedure to *Behavioral and Brain Sciences*, the information for which is shown in Table 7.4. In this case the calculation is as follows:

$$\text{Five-year IF}_{\textit{Behav Brain Sci}} = 1{,}049/64 = 16.391$$

It can be seen that in both cases the five-year impact factor has a lower value than that obtained by the traditional method. However, this difference is much greater in the case of *CA: A Cancer Journal for Clinicians* (69.026 for the two-year impact factor and 45.611 for the five-year impact factor). In contrast, the difference for *Behavioral and Brain Sciences* is quite small (17.462 and 16.391, respectively). As noted previously, this is due to the way in which citations behave in different research fields.

Finally, let us consider the two typical citation behaviours that appear in scientific literature. Instead of a single journal, we will analyse the discrepancy between the two-year and five-year impact factors for journals from the same *subject category*. The first example concerns the subject category *Medicine, General and Internal* from the 2007 Science

Table 7.4 **Citations received and items published in *Behavioral and Brain Sciences* from 2002 to 2006 (2007 JCR)**

	2006	2005	2004	2003	2002	Total
Citations in 2007 to items published in	98	356	201	196	198	1,049
Number of items published in	12	14	14	12	12	64

Citation Index, and focuses on the ten journals that have received the highest impact factors in this category (Table 7.5).

Most of the journals shown in this table obtain a higher impact factor when this is calculated using the traditional two-year target window rather than the five-year one. Of course, there are exceptions and three of the journals shown obtain a higher impact factor when considering a longer time period.

This pattern is typical of those disciplines that advance quickly and, therefore, it makes perfect sense that their articles are cited more quickly but may then lose some citation impact after the first few years.

Now let us look at the ten journals which have received the highest traditional impact factors for the research area of the ISI Web of Knowledge that includes bibliometric studies (Table 7.6). This subject category is called *Information Science and Library Science*. In this case, all ten journals in this category obtain a higher impact factor when this is calculated for a five-year rather than a two-year period. This is the characteristic pattern of disciplines that require more time to fully realise their impact.

Occasionally, impact factor calculation may be interrupted, for example when a journal changes its name. In this case, an approximate and more realistic value of the impact factor for the period of time following the journal's change of name is calculated. The so-called *combined impact factor* is often applied in these cases.

Table 7.5 **Two- and five-year impact factors of ten journals belonging to the subject category *Medicine, General and Internal* from the Science Citation Index (2007 JCR)**

Abbreviated journal title	2-year IF	5-year IF
New Engl J Med	52.589	45.941
Lancet	28.638	24.201
JAMA – J Am Med Assoc	25.547	25.793
Ann Intern Med	15.516	14.913
Annu Rev Med	13.415	11.358
Plos Med	12.601	21.727
Brit Med J	9.723	9.069
Arch Intern Med	8.391	8.570
Can Med Assoc J	7.067	6.735
Ann Med	5.779	4.615

Table 7.6 **Two- and five-year impact factors of ten journals belonging to the subject category *Information Science and Library Science* from the Social Sciences Citation Index (2007 JCR)**

Abbreviated journal title	2-year IF	5-year IF
Mis Quart	5.826	9.257
J Am Med Inform Assn	3.094	3.489
Inform Syst Res	2.682	6.579
Annu Rev Inform Sci	1.963	2.810
J Manage Inform Syst	1.867	3.229
J Health Commun	1.836	2.021
Int K Geogr Inf Sci	1.822	2.068
Inform Manage-Amster	1.631	2.756
J Inf Technol	1.605	2.045
Inform Syst J	1.531	2.085

In order to understand how the combined impact factor is calculated let us look at a journal that did change its name. *Obesity Research*, the official journal of the Obesity Society, was founded in 1993 but then changed its name to *Obesity* in 2006. As the impact factor calculation is based on the citations given to a journal in the two years prior to the year selected, there will be an imbalance in the journal's impact factor in the years immediately after the name change.

For this example we will consider the 2007 Journal Citation Reports, the latest edition available at present, to analyse how the impact factor has changed as a result of the name change from *Obesity Research* to *Obesity*. The citations received to articles published in *Obesity Research* as well as the number of items published from 2005 to 2006 are shown in Table 7.7.

As can be seen in the table, this journal published a total of 255 items during 2005, but none were published during 2006. Obviously, this is because the journal changed its name in 2006 and so all its articles for this year were published under the new name. Using the information shown in Table 7.7 it is possible to obtain the impact factor for *Obesity Research* by means of the following formula:

$$\text{Impact factor}_{Obes\ Res} = 1{,}263/255 = 4.953$$

Table 7.7 **Citations received and items published in *Obesity Research* (2007 JCR)**

	2006	2005	Total
Citations in 2007 to items published in	58	1,205	1,263
Number of items published in	0	255	255

Additionally, we can gather the number of citations and items published by the journal *Obesity*, as shown in Table 7.8. In this case there is a complementary distribution of items and references as compared with its former name. Obviously, no items were published by this journal in 2005 as it did not yet exist. In addition, the number of citations received to items published in 2005 has been underestimated, as only two citations were found. Given that the journal *Obesity* only appeared in 2006 it might seem surprising that two citations to articles from this journal were found. However, there is a plausible explanation for this fact. Once a journal has changed its name all the articles published under its former name will belong to the new journal title. Therefore, it is possible that some articles have been cited with the new name of the journal instead of its former title.

Thus the impact factor for this journal in 2007 is calculated as follows:

$$\text{Impact factor}_{Obes} = 477/294 = 1.520$$

As we can see, this journal has two impact factors, one under the name of *Obesity Research* and another corresponding to its new name *Obesity*. The decrease in the impact factor from the first (4.9) to the second (1.5) is also significant and is due to the number of items or citations included in the analysis. At all events, these citations and items belong to the same journal.

An alternative way of compensating for this effect is to calculate the combined impact factor, which requires gathering all the citations received and items published by *Obesity Research* and *Obesity* in order to obtain a single impact factor for 2007. Table 7.9 shows the values needed to obtain the combined impact factor for these journals.

Table 7.8 **Citations received and items published in *Obesity* (2007 JCR)**

	2006	2005	Total
Citations in 2007 to items published in	445	2	447
Number of items published in	294	0	294

Table 7.9 **Combined impact factor for *Obesity Research* and *Obesity* (2007 JCR)**

	2006	2005	Total
Citations in 2007 to items published in:			
Obesity Research	58	1,205	1,263
Obesity	445	2	447
Total			**1,710**
Number of items published in:			
Obesity Research	0	255	255
Obesity	294	0	294
Total			**549**

The combined impact factor for these journals is calculated by obtaining the total number of citations received during 2005 and 2006 in both journals, divided by the total items published in both journals. The combined impact factor is thus as follows:

$$\text{Combined impact factor}_{Obes\ Res\ -\ Obes} = 1{,}710/549 = 3.115$$

The combined impact factor is more accurate for cases in which a journal has made a formal change, such as a name change, since some citations and items published may then be underestimated for a given period of time. In this example it seems that the new journal *Obesity* had significantly decreased its impact factor due to the name change, but this was only due to the nature of the calculation. Calculating the combined impact factor yielded a value that seems to explain more accurately the real impact that these publications have in the scientific community.

The procedure used to calculate the combined impact factor can also be applied to obtain the *five-year combined impact factor*. Continuing with the same example we will now consider the citations received and the items published by the journals *Obesity Research* and *Obesity* in the last five years. As before, the aim of this analysis is to obtain a more accurate impact factor for the journal's new title, since the count of citations and items is underestimated before 2006. Once again there is a significant difference between the five-year impact factor for the journal *Obesity Research* and that for *Obesity*. While the former has a five-year impact factor of 5.353, the latter has a value of 1.544. These values can be obtained by means of the procedure used to calculate the five-year impact factor shown previously, and with the number of citations and items published according to the Journal Citation Reports. Table 7.10 gathers

Table 7.10 Five-year combined impact factor for *Obesity Research* and *Obesity* (2007 JCR)

	2006	2005	2004	2003	2002	Total
Citations in 2007 to items published in:						
Obesity Research	58	1,205	1,486	1,170	904	4,823
Obesity	445	2	7	0	0	454
Total						**5,277**
Number of items published in:						
Obesity Research	0	255	262	201	183	901
Obesity	294	0	0	0	0	294
Total						**1,195**

all the citations and items found for both journals over the last five years available in the 2007 Journal Citation Reports (i.e. from 2002 to 2006).

The information shown in this table can now be used to calculate the more accurate, combined five-year impact factor for these two journals. This five-year impact factor is calculated by dividing the total number of citations received by both journals during this period of time by the total number of items published by the two journals. The following formula shows how to derive the combined five-year impact factor:

$$\text{Five-year combined IF}_{Obes\ Res\ -\ Obes} = 5{,}277/1{,}195 = 4.416$$

As before, we thus obtain a single five-year impact factor for 2007 that gathers data for both these journals.

The combined analysis only makes sense in the years immediately following the name change. Once the count of citations and items ceases to overlap between the two journal names, the impact factor can be obtained by means of its original formula.

Having looked at how to estimate the impact a journal has within the scientific community, we can now move on to consider an index that assesses whether the impact of publications in a country in a given research field is compatible with its research efforts. The *publication efficiency index* (PEI) is based on an index formulation developed by Frame (1977) and later used by Garg (2002) and Guan and Ma (2004, 2007).

A PEI value higher than 1 means that the impact of publications in this country is greater than the research effort made. The analysis is based on the ratio of citations received per item published by a country compared to this ratio for all the countries included in the analysis.

The general formulation of the PEI is as follows:

$$PEI = \frac{TNC_i / TNC_t}{TNP_i / TNP_t}$$

where TNC_i is the total number of citations received by journal i, TNC_t is the total number of citations received by all countries, TNP_i is the total number of items published by country i, and TNP_t is the total number of papers published by all countries.

Let us consider an example of how the PEI can be applied in a given research field, in this case *obesity surgery* in the Web of Science during 2007. The data gathered for this area of research over a short period of time will be sufficient to illustrate this index. However, its application to larger databases would provide a more exhaustive analysis in broader research fields and over longer periods of time.

To carry out this analysis we first need to gather information about the number of items published by each country, as well as the number of citations they received. Table 7.11 shows this information.

It can be seen that authors from 18 countries have published articles about obesity surgery during 2007. The first and second columns show the number of items published by these countries in the period selected, as well as the number of times that these items have been cited.

In those countries where the PEI value is higher than 1, the impact of their publications is also higher than the research effort made. In contrast, a PEI value lower than 1 means that despite the research efforts made by the country the impact of its research is very weak.

In this example, countries are ranked by the number of items published. However, the most productive countries are not necessarily those which obtain higher PEI values. As we can see, Austria is the country with the highest PEI value, corresponding to 3.98. This means that Austria has made an effort to publish in this area, and suggests that its three articles have been well accepted and recognised by other authors in this field as they have received a total of 36 citations. In fact, the highest frequency of citations for the studies gathered in this example corresponds to one of the articles published by Austria, which obtained all 36 citations. It is clear, therefore, that the impact of this publication has been higher than the efforts needed to publish it, and the other articles published by Austria have not been cited. Of course, there are other countries whose articles have had a good impact, for example Argentina, Italy, France, the USA and Switzerland, all of which obtained PEI values higher than 1. Although their values appear moderate with

Table 7.11 Publication Efficiency Index (PEI) in *obesity surgery* (2007 WoS)

Country	Items published	Times cited	PEI
USA	34	115	1.12
France	9	31	1.14
Brazil	6	9	0.50
Germany	6	10	0.55
Italy	5	25	1.66
UK	4	4	0.33
Spain	4	1	0.08
Australia	3	8	0.89
Austria	3	36	3.98
Taiwan	3	4	0.44
Sweden	2	5	0.83
Argentina	1	5	1.66
Ecuador	1	1	0.33
Netherlands	1	1	0.33
Philippines	1	1	0.33
Poland	1	0	0.00
Portugal	1	0	0.00
Switzerland	1	3	1.00

respect to Austria, their publications have still obtained more impact than the efforts required in publishing them. The remaining countries shown in this table obtain smaller indices, and there are even two countries (Poland and Portugal) with PEI values equal to 0. This means that despite their publishing efforts, their work has so far had no impact within the scientific community, as their only work has not been cited.

This index can be used to compare the impact of articles published by different countries. However, as with all the other indices described above, the PEI has its limitations. Indeed, comparisons between countries will evidently offer a clear advantage to some of them as there is no level playing field when it comes to conducting research which depends on many factors. Thus some countries will be in a much stronger position as regards the publication of articles, and this greater productivity increases the likelihood of obtaining citations. Once again, citations have to be treated carefully, and the interpretation of indices based on citations must be conservative.

Journal h-index

This section looks at how to use the h-index (Hirsch, 2005) to assess a journal's scientific output. As we have seen previously, this index was initially designed to assess an individual's scientific output, considering both the productivity and the impact of a researcher's publications. Recall that a scientist has an h-index if h of his or her N papers have at least h citations each and the other $(N - h)$ papers have $\leq h$ citations each.

However, in addition to individual productivity the h-index has also been applied to other sources of data such as topics (Banks, 2006), library management (Liu and Rousseau, 2007) or different levels of productivity by means of the successive h-index (Egghe, 2008a).

For now let us focus on the proposal of Braun et al. (2006), who applied the h-index to assess journal impact. The journal impact factor developed by Garfield (1955) and Garfield and Sher (1963) is now widely accepted to be the way to assess a journal's impact factor. However, Braun et al. propose the h-index as a supplement to the evaluation of journal impact. Furthermore, they highlight a number of advantages of the h-index over the traditional impact factor. First, they point out that the h-index is more robust, as it is not sensitive to an excess of uncited papers or to highly cited ones. Moreover, they regard as positive the fact that the h-index combines the number of publications with the citation rate in a balanced way that will reduce the overrating of some journals.

The procedure to calculate the h-index here is the same as for author assessment (see Chapter 6 on the Hirsch index), except for the period of time selected. In the case of an author, all the papers published during the researcher's career are included in the ranking to derive the h-index. However, in the case of journals not all the documents published are included, but only those published in a given period. In the simplest case, only papers published during a single year are taken into account (Braun et al., 2006).

Once we have selected the journal and the period of time to be included in the h-index analysis, the papers published during this period must be ranked in decreasing order according to the number of citations received. This information can be easily obtained directly from the ISI Web of Science. The h-index can then be calculated by identifying how many h papers have at least h citations each, while the other $(N - h)$ papers have received $\leq h$ citations each. Furthermore, by gathering the h-indices of certain journals belonging to the same research area we can rank them according to their h-index.

Let us look at an example of ranking journals according to their h-indices. Since Braun et al. (2006) propose this index as a supplementary assessment to the traditional impact factor, both the h-index and the impact factor will be included in this ranking.

First, the period of time in which the journal will be assessed needs to be fixed, and then a group of journals from the same research area must be selected. It is advisable to choose journals from the same research field due to differences in the h-index between fields (Hirsch, 2005). The present example focuses on the year 2003. As one of the components of the h-index is the number of citations received by a journal, it is necessary to gather papers published some time ago, so that they have time to be cited. Articles published in 2003 would have had enough time (in this case six years) to be cited. Once the time period has been decided the next step is to select the journals whose h-indices will be calculated. In this case, we will select the 20 journals from the Science Citation Index that had the highest impact factor values in the Journal Citation Reports. Since the impact factor and the h-index for journals are complementary impact indices, we will be able to compare their values for the journals selected. As the h-index will be calculated according to articles published during 2003, the impact factor will also be obtained from the 2003 Journal Citation Reports. In this case, we are simply comparing the journal rankings according to their impact factor and h-indices. However, if we wish to compare the h-indices between journals a more selective search must be carried out in order to gather only comparable journals from the same subject category.

Table 7.12 shows the journals selected, along with the h-indices and the impact factor values and ranks. This table shows the 20 journals with the highest impact factor values in the 2003 Journal Citation Reports. However, they are ranked according to their h-index. When two journals have the same h-index they are ranked in the same position. The table also shows the total number of documents published by the journal during the period selected. The number of publications is important in the h-index calculation as it represents the maximum value that *h* can reach.

It is important to mention that in this bibliographic search of articles published in these journals, the type of document has also been taken into consideration. This is because the impact factor calculation is based on the citations received by only a certain type of document, mainly research articles and reviews. Consequently, the h-index for these journals has also been calculated according to these documents so as to be comparable to the impact factor. Had we taken into consideration for the h-index calculation all the documents published by a journal, the resulting h-index

Table 7.12 h-rank of the top 20 impact factor journals from the Science Citation Index (2003 JCR)

h-rank	Abbreviated journal title	Documents	h-index	Impact factor	IF rank
1	*Science*	926	194	29.781	11
2	*New Engl J Med*	366	153	34.833	5
3	*Cell*	281	119	26.626	17
4	*Nat Genet*	182	97	26.494	18
5	*Nat Med*	167	90	30.550	9
6	*Nat Immunol*	141	80	28.180	12
7	*Nat Rev Mol Cell Bio*	89	60	35.041	4
8	*Nat Rev Immunol*	72	59	26.957	16
9	*Nat Rev Neurosci*	82	56	27.007	15
10	*Nat Rev Cancer*	61	55	33.954	6
11	*Nat Rev Genet*	84	50	25.664	19
12	*Nature*	856	46	30.979	8
13	*Physiol Rev*	33	32	36.831	3
13	*Rev Mod Phys*	28	32	28.172	13
14	*Annu Rev Cell Dev Bi*	27	27	22.638	20
15	*Pharmacol Rev*	26	25	27.067	14
16	*Annu Rev Immunol*	25	24	52.280	1
17	*Annu Rev Biochem*	24	22	37.647	2
17	*Annu Rev Neurosci*	23	22	30.167	10
18	*CA - Cancer J Clin*	16	13	33.056	7

may have been overestimated in comparison to the impact factor. At all events, it should be noted that there are many incongruities in the document count. By knowing the number of documents included in the impact factor calculation it can be seen that they do not always coincide with the number of substantive articles given by the bibliographic count conducted through the ISI Web of Science. However, if impact factor is based on the data provided in the ISI Web of Science, these values should be the same. This may be due to the inclusion of certain types of document in the count. In this example the number of documents obtained refers to those found in the bibliographic search, which corresponds to the highest value that the h-index can reach. However, in nine of the 20 articles, incongruities in the document count were found.

In the table it can be seen that only one journal reaches the maximum value attainable by the h-index. This journal, *Annual Review of Cell and Developmental Biology*, published 27 papers in the year 2003, all of which obtained 27 or more citations each. Since the *h* value and the impact factor are complementary impact indices, the table also shows the ranking of journals according to their impact factor. It can be seen that the positions in the *h* and impact factor rankings do not necessarily coincide. For example, the journal *Science* obtained the highest h-index, but was ranked eleventh according to its impact factor. In this case, the h-index may have reached a higher value due to the large number of documents published by this journal. As such, the scientific output and impact of this journal is better valued when applying the h-index rather than the impact factor. In contrast, the journal *Annual Review of Immunology*, which had the highest impact factor, was only ranked sixteenth according to its h-index. In this case, the small number of documents published (25) would have an important influence on the values obtained. As this journal has published a small number of documents compared with the hundreds of documents published in the same period by others, the highest value that the h-index could reach will also be small. In contrast, this small number of published documents increases the likelihood of the journal obtaining a higher impact factor, because a smaller denominator will be considered.

Similar results were found by Braun et al. (2006) when comparing the h-index and the impact factor of a group of journals. In line with the present results they found differences in the h-index and impact factor rankings of the same journals. Specifically, they report that some journals, despite being prestigious in their field, were not ranked among the top impact factor journals. These will be journals that receive a significant number of citations for their papers and, as such, high h-indices, but they will have small impact factor indices compared with the other journals in their field. In conclusion, the h-index is a complementary index for assessing journal impact as it gives a different point of view within journal citation analysis.

The promise of this index when applied to journal assessment has been stated by Braun et al. (2006), who consider that much systematic analysis and statistical background work will be possible in the future.

Finally, it is worth mentioning the comparison between the impact factor and the h-index. Both are considered to be quality measures for evaluation purposes. As we have seen, the impact factor has been applied to journals while the h-index has mainly been applied to authors, although also to evaluate journal productivity and impact. Nowadays, there is a need to evaluate individual performance (Jeang, 2007), but this must be complemented by an evaluation of journal quality (Hönekopp and Kleber, 2008). Given that scientist and journal evaluation forms part of the overall assessment of the scientific community, Hönekopp and Kleber (2008) tested whether the journal impact factor or the author h-index offers the best prediction of the total citation count. They start from the idea that the total citation count can be used as a measure of an article's quality (Hirsch, 2007; Simonton 1997, 2003). Their analyses were based on a bibliographic search of papers in which the editorial board members of the journal *Retrovirology* had participated as the first or last author. The search was conducted by means of Google Scholar and covered a period of five years. The reason for including participation as both first and last author is that Hönekopp and Kleber considered that the board members of this journal may have reached a stage in their career in which the papers most representative of their work are those for which they are the last author. They correlated the journal impact factor and the author h-index with the total number of citations and found that the impact factor predicted the total number of citations. However, no correlation was found between the author h-index and the total number of citations. According to Hönekopp and Kleber, the impact factor outshines the h-index when predicting the future number of citations of an article.

Journal self-citations

Self-citations often represent a significant percentage of the citations that a journal receives. In fact, it is estimated that approximately 13 per cent of the total number of citations that a journal receives are self-citations (Thomson Reuters, 1994). The fact of a journal citing itself will have direct consequences on the value of its impact factor, hence the debate over whether to include self-citations in the count of a journal impact factor. As we saw in a previous section, the first formulation of the journal impact factor (Gross and Gross, 1927) did not include self-citations in the count. This was because the authors considered that self-citations would have an important effect on the citation count used to calculate impact factor and lead to its value being overestimated. In these early years of research on the impact factor, other authors also discussed the role of self-citations in this index and likewise decided that self-citations should be excluded (Allen, 1929; McNeelly and Crosno, 1930; Gregory, 1937; Westbrook, 1960). However, the impact factor formulation developed by Garfield and Sher (1963) did include self-citations in the analysis and this formulation is still used at present.

The behaviour of journal self-citations over time has also been an object of study. Biglu (2008) carried out a self-citation study of journals randomly selected from the Journal Citation Reports for the period 2000 to 2005. He found that the number of self-citations increased over this period, as did the total number of citations. Of course, this implied an increase in the journal impact factor over time. However, when taking into consideration self-citation rates rather than the absolute count of citations, there was a clear difference between journals with a high or low rank (in terms of the number of citations). This study demonstrated, for a sample of 500 journals, that those with high citation rates had a lower self-citation rate (2 per cent) than did the low-ranked journals (17 per cent).

Arguments in favour of excluding self-citations from impact factor analysis have started to have an effect. For example, recent improvements have been made to the Journal Citation Reports (Thomson Reuters, 2009), not only in terms of the five-year impact factor but also regarding self-citation analysis. At present, the total count of self-citations made by a journal in all years is shown. The number of self-citations included in the analysis of the two-year impact factor is also included, as well as an *impact factor calculation excluding self-citations*. As such, two impact factors are given and their value including or excluding self-citations can be compared.

By way of an example let us consider the influence that self-citations have on the impact factor calculation for the *Journal of Applied Psychology* using information provided by the 2007 Journal Citation Reports. As before, we first need to know the number of citations that this journal has received during 2007 to the previous two years (2005 and 2006), as well as the number of items published by this journal during these two years. Since the aim is to determine the influence of self-citations on the impact factor, we also require the number of self-citations that this journal has received. Table 7.13 gathers all the information needed to calculate the impact factor, both including and excluding self-citations.

The impact factor is given by dividing the number of recent citations by the number of recent items (Table 7.13). Obviously, the number of total and recent citations will be smaller when excluding self-citations and, consequently, the corresponding impact factor will be also smaller. For the case of the *Journal of Applied Psychology*, the impact factor decreases significantly (from 3.047 to 2.425) when excluding self-citations. As we can see, 13 per cent of the total number of citations (1,477 of 11,182) are self-citations, and this proportion is even higher in the case of self-citations

Table 7.13 **Self-citations analysis in the *Journal of Applied Psychology* (2007 JCR)**

	Including self-citations	Excluding self-citations
Total citations	11,182	9,705
Number of recent citations	652	519
Number of recent items	214	214
Impact factor	3.047	2.425

included in the impact factor calculation. Here 20 per cent (133 of 652) of the citations received in 2007 to items published during the previous two years were self-citations. This illustrates that self-citations play an important role in the impact factor calculation by increasing its value. One way to avoid this inflation of impact factor is therefore to exclude self-citations from the calculation.

There is another way in which self-citations may distort the impact factor calculation. As will be seen in the next section, there is a group of journals referred to as *cited-only journals*. The defining characteristic of these journals is that their references are not available in the Journal Citation Reports database. However, it is possible to count the number of citations they receive by other journals. If we are comparing journals by means of the impact factor that includes self-references, these journals will be at a disadvantage, for if their references are not included, neither will their self-citations.

The decision to analyse the impact factor while excluding self-citations is a great achievement for the current system of journal evaluation. As the traditional impact factor is widely used to assess journals, authors or even to distribute resources in terms of grants or research projects, it has important repercussions in daily practice. Consequently, and as Archambault and Larivière (2009) state, the traditional impact factor can be a potent way for editors to manipulate their journal's standing by urging authors to cite the journal in which they publish. If self-citations are counted it is also possible to increase the impact factor of a journal by encouraging authors to cite papers from the journal in which they are trying to publish. This bad practice could be reduced if self-citations did not have the importance they currently have.

In what follows we will look at the extent to which the scientific community has accepted an impact factor that excludes self-citations. It will be argued that better praxis in the field will only be achieved by using such a criterion for journal impact assessment.

Journal cited and citing half-life

This section looks at two indices that consider the years of publication of citations. The difference between them rests on the source of the data included in the analysis. The first of these indices (*cited half-life*) focuses on citations received by a given journal, while the second (*citing half-life*) considers the references included in the articles published by a journal.

The cited half-life index is based on the references received by a journal and computed in the Journal Citation Reports of the ISI Web of Knowledge (i.e. in the Science Citation Index and Social Sciences Citation Index). Therefore only those journals included in this database will be included in this analysis. This index was developed to provide an indicator as to the long-term value of source items in a single journal publication.

The index represents the median age of the citations given to a journal in a given year and is used to estimate the impact of a journal. The cited half-life is the number of years, counting back from the current year, that account for 50 per cent of the total citations received by the journal in the current year.

Only those journals that have received 100 or more citations in the JCR during a given year have a cited half-life. A higher or lower value for this index does not imply any particular value for the journal, but simply indicates the age of the references it received in the current year.

An example of cited half-life is that of the journal *Experimental Brain Research*, which obtained a cited half-life of 9.5 years when considering citations received in 2007. This means that 50 per cent of the references given to this journal in 2007 belong to the last 9.5 years. This journal was founded in 1966 and has published more than 11,000 articles. The cited half-life provides information about how recent the citations received by this journal are. Since half the citations belong to the last 9.5 years they correspond to articles published in *Experimental Brain Research* between approximately 1998 and 2007. The remaining citations refer to articles published in this journal before 1998 (i.e. from 1966 to 1997).

Another example of cited half-life is that obtained by the journal *Scientist*, which corresponds to 2.8 years. This means that citations made to this journal are quite recent, since half of them belong to the last 2.8 years. Given that our data come from the 2007 Journal Citation Reports, this index means that half the citations received by the journal *Scientist* in this database correspond to articles published from around 2005 to 2007. It would thus seem that the newer articles published by this journal are of interest to other researchers in its field, and it can be stated that this journal provides rapid communication of current information.

Of course, the cited half-life can be influenced by the age of the journal. Those journals which have been publishing for longer have a greater likelihood of receiving citations to their oldest articles. However, this does not mean that the oldest journals will obtain higher cited half-life indices, since this index depends on the citations received in a given year. In contrast, a younger journal will never obtain a higher cited half-life index. An example of this phenomenon is shown in Table 7.14, which shows the citations received by two journals during 2007. The first of these is the previously analysed *Scientist*, which was founded in 1986, while the second is the *Annual Review of Clinical Psychology*. The latter journal is very new compared to the other since it was founded in 2005.

Although these two journals have obviously been publishing articles for different lengths of time the cited half-life of their articles in 2007 is quite similar: 2.8 for *Scientist* and 2.4 for *Annual Review of Clinical Psychology*. Thus they both received half of their citations in a similar period of time, more or less since 2005. However, the other half of the citations will be distributed differently. While in the first journal the remaining citations will be distributed over a long time period (1986 to 2005), in the second all these citations will be concentrated into a brief period of time, basically 2005. Obviously, the point at which a journal is founded constitutes a boundary to this index and establishes the maximum value which the cited half-life can take. However, this does not imply a higher or lower value for this index, as it will depend on the citations it receives.

Cited half-life takes into consideration the frequency of citations received in the ten years prior to the year analysed. Another example of the data involved in cited half-life analysis is that shown in Table 7.15 which gives the frequency of citations received by the journal *Experimental Brain Research* during the year 2007. Despite the long career of this journal, only those citations received in the last ten years are available. As can be seen in this table, half the citations (50 per cent) are located around the years 1998 and 1999, since the cumulative percentage of citations received is shown.

The cited half-life indices that are higher than 10 are not specified. In this case, the journal is said to have a cited half-life > 10 years, but the exact age of half of its citations is not specified. This means that citations received by this journal do not belong to current contents published in the journal. This may be a characteristic of journals that publish classic studies which continue to be cited as they represent the basis of a research field.

However, the journals included in the Journal Citation Reports are only a proportion of the journals publishing in a given area. Thus these indices

Table 7.14 Citations received in 2007 by the journals *Science* and *Annual Review of Clinical Psychology*

	2007	**2006**	**2005**	**2004**	**2003**	**2002**	**2001**	**2000**	**1999**	**1998**	**1997-all**	Total
Scientist												
Citations from 2007	42	53	56	40	24	14	6	7	8	4	29	283
Cumulative %	14.84	33.57	53.36	67.49	75.97	80.92	83.04	85.51	88.34	89.75	100.00	
Annu Rev Clin Psycho												
Citations from 2007	7	21	145	1	0	0	0	0	0	0	0	174
Cumulative %	4.02	16.09	99.43	100.00	100.00	100.00	100.00	100.00	100.00	100.00	100.00	

Table 7.15 Citations received by the journal *Experimental Brain Research* in 2007

	2007	**2006**	**2005**	**2004**	**2003**	**2002**	**2001**	**2000**	**1999**	**1998**	**1997-all**
Citations from 2007	153	867	1,115	927	1,076	1,100	940	1,017	1,088	891	8,237
Cumulative %	0.88	5.86	12.26	17.59	23.77	30.08	35.48	41.32	47.57	52.69	100

may be biased by the citation count as it only takes into consideration citations made by the other journals included in this database.

The practical implications of the cited half-life of journals have been discussed by Ladwig and Sommese (2005), who point out that this index may be useful in assisting collection management and archiving decisions. These authors developed a statistical model based on the cited half-life in order to manage cancellation subscriptions to online journals. Specifically, they sought to improve the decision-making process carried out by the university libraries of their city. The traditional method used by the university to decide whether to continue with or cancel the subscription to an online journal was based on a rule of cost-efficacy. This meant that the subscription cost of a journal was divided by the number of full-text downloads for one year in order to calculate the cost per download. In the event that the cost per download was higher than the cost of obtaining an article from this journal through another source, the article subscription was cancelled. The model proposed by these authors includes the cited half-life of the article in the decision-making process, as they considered that the number of downloads should be adjusted for the journal's half-life. Then, instead of taking into consideration the number of downloads made for the year of subscription, it would be necessary to count how many downloads there have been of articles published in the journal in the previous years. Their model was based on the idea that download counts would be reasonable for short cited half-life journals, but would produce a significant undercount for long cited half-life journals. Of course, the availability of an online paper will be influenced by the recent availability of electronic journals, and the fact that articles from long cited half-life journals would be obtained from other sources, as they would not be available online.

Another index applied to journal citation analysis concerns the journal's citing half-life. As before, the index is based on the references made by a given journal and, similarly to the previous index, the citing half-life of a journal is the median age of the items that the journal cites in a given year. Thus half of the citations or references made in the journal in a given year will be included within the citing half-year. Once again, only those journals that contain more than 100 references will have a citing half-life index. Let us consider this index by looking at the same journals used in the example of cited half-life. Of course, the two indices do not need to be similar as one is based on citations received whereas the other is based on the references made in a journal.

In the case of *Experimental Brain Research*, its citing half-life is nine years. This means that half the total number of references included in the articles published during 2007 were from articles published over a period of nine years (from 1999 to 2007). In the case of the journal *Scientist*, the citing half-life is 2.4 years. Thus half of the references made in the articles published during 2007 in this journal refer to quite recent works, published between 2005 and 2007.

The procedure used to calculate the citing half-life is the same as that for the cited half-life index but from a different point of view, i.e. considering the references made in the journal.

There is one specific case in which the citing half-life cannot be obtained, namely for what are known as *cited-only journals*. This is a group of journals from the Journal Citation Reports of the ISI Web of Knowledge whose references are not included in the database. Consequently, these journals do not have a citing half-life as their references are not available for inclusion in the analysis.

As mentioned in the previous section, the fact that references from cited-only journals are not included in the database has implications for other analyses used to assess journals. The first consequence concerns the count of self-citations. As discussed above, self-citations often represent a significant percentage of the citations received by a journal. In the case of cited-only journals, self-citations are not taken into consideration in any of the analyses, since references from these journals are not available in the database. Consequently, other analyses, such as the impact factor, will also be biased, as the recount of citations will be incomplete for this group of journals.

Other indices that can be obtained by taking into consideration all the journals belonging to a given subject category are what are termed the *aggregated cited* and *citing half-life*. The procedure to obtain these indices is the same as that for the journal's cited and citing half-life. However, in this case we consider all the articles published in all the journals belonging to a subject area. Thus these indices are obtained as the median age of the cited and citing half-lives of all the journals from this area. For instance, if we consider the subject *Neurosciences* from the 2007 Journal Citation Reports, we find a total of 211 journals belonging to this category, and the median of the cited half-life values of all these journals gives an aggregated index of 6.8 years. This aggregated index gives an idea of the turnover rate of the body of work in this subject area. It is also possible to obtain the aggregated citing half-life of this category, which for *Neurosciences* is 7.5 years. However, in this case we are considering all the references made by the articles of these 211 journals belonging to this subject category. The index value of 7.5 years means that half the references made by these articles belong to articles published between 2000 and 2007.

The cited and citing half-lives can be related to the *ageing process* of publications. As Moed et al. (1998) explain, the ageing process of a journal can be viewed as a combination of two processes: maturation and decline. In the years after a paper's publication the number of citations received will increase up to a maximum citation rate. This maturing phase will then be followed by a decrease in the number of citations received, which will represent a decline. In this context, cited and citing half-lives will be indicators of the ageing process of a journal.

Specifically, Tsay (2009) points out that the cited and citing half-lives can be related to the obsolescence of a journal, which will be associated with its decline (Moed et al., 1998). Obsolescence can also be useful when assessing the value of a journal. What is of interest in this case is how long a publication continues to be used after it has been published. When an article is no longer used it can be said to be obsolete. The same idea can be applied to a journal, where the rate of obsolescence would be determined by the cited and citing half-lives.

If obsolescence is related to these indices which are, in turn, influenced by productivity, it can be assumed that obsolescence is influenced by journal productivity. This suggests that the productivity of a journal will be a significant factor in the calculation of other indices, as the more productive journal will also receive more citations. Wallace (1985) showed that the most highly productive journals had lower journal cited half-lives. Furthermore, articles with higher cited half-lives were less productive in terms of the article count. In the same line, the study by de Queiroz and Lancaster (1981) found that literature in their chosen field of research was growing while the median age of the citations received was decreasing. Thus we can state that since the obsolescence of a publication will be determined by the cited half-life, it will also be influenced by productivity.

As before, the validity of these indices is the subject of debate in the scientific community, as many authors consider their application and interpretation to be dubious. In Chapter 8 we will review the main arguments for and against the cited and citing half-lives of journals.

Eigenfactor metrics

The Eigenfactor metrics (Bergstrom, 2007; Bergstrom et al., 2008) include two measures based on citation data to assess the influence of a journal in relation to other journals. These two measures are the *Eigenfactor score* and the *Article Influence score*. Both have recently been included in the Journal Citation Reports and are available for 2007 and later. Thomson Reuters (2009) argues that the inclusion of these measures in the JCR provides a broader perspective on journal influence through specific measures that are now widely accepted by the scholarly community.

Eigenfactor metrics are based on the idea that connections in scientific literature are made by citations, which reflect the researcher's view about which papers are relevant to his or her work. The aim of developing a new measure of citation counts was to improve upon the simple citation counts used to measure scientific influence (Bergstrom, 2007), as well as to complement other measures based on citation counts. As Bergstrom and West (2008) state, these measures have been proposed to remedy two biases associated with impact factor: the failure to take into account, firstly, the differences in prestige between citing journals and, secondly, the differences in citation patterns across disciplines.

Eigenfactor measures rank journals to identify the most influential, i.e. those that will receive more citations from other influential journals. Thus they are based on the idea that a single citation from a high-quality journal may be more valuable than multiple citations from secondary journals (Bergstrom, 2007). This is an improvement introduced by Eigenfactor metrics, since other analyses based on citations do not account for where citations come from (Bergstrom et al., 2008). As such, a citation made by a review article that includes a large number of references will count less than a citation made by a research article that only cites papers which are relevant to its content (Bergstrom, 2007).

Another important contribution of Eigenfactor measures is that they are adjusted for citation differences across disciplines. By using the whole citation network these measures account for differences in citation patterns between fields, thus enabling a better comparison across research areas (Bergstrom, 2009).

Another characteristic of Eigenfactor metrics is that they take into account the citations received by journals during the previous five years. The reason for choosing this extended time span is that in many research areas, articles are more commonly cited for a certain period after their publication (Bergstrom, 2009).

These indices have been calculated for the publications indexed in the Journal Citation Reports, as well as for other documents such as books, newspapers or PhD theses. They therefore provide the opportunity to assess the importance and influence of documents other than journals in the scholarly literature (Bergstrom, 2007). Finally, it is important to note that Eigenfactor measures are freely available (Bergstrom, 2009).

The Eigenfactor score measures the total influence of a journal on the literature, estimating how often a journal is used by researchers. Specifically, it represents the total collective value provided by all the articles published in a journal in one year (Thomson Reuters, 2009). This measure is based on a simple model of research in which readers follow chains of citations moving from one journal to another (Bergstrom, 2007). When a researcher reads an article, he or she will probably be interested in some of the references it includes, and may thus proceed to locate and read a given article that has been published in another journal. This procedure can be repeated several times by the reader. This is the idea underlying the mathematical algorithm of the Eigenfactor score, which uses the structure of the entire network to evaluate the importance of each journal, cutting across all disciplines (Thomson Reuters, 2009). The Eigenfactor score counts citations to journals in both the sciences and social sciences, excludes journal self-citations and takes into consideration a five-year period of citation activity. The size of a journal will influence this measure, since larger journals will have more citations and will be consulted more often by researchers, which leads to higher Eigenfactor scores (Bergstrom, 2007). This means that the measure scales with the size of a journal and then measures the journal's total importance (Bergstrom and West, 2008).

The Article Influence score reflects a journal's prestige (Bergstrom and West, 2008). In contrast to the Eigenfactor measure, which seeks to estimate the importance of an article according to the journal in which it is published, here the size of the journal is not relevant. What is important when measuring the prestige of a journal is the average influence of the articles it publishes (Bergstrom and West, 2008). Comparable to the Eigenfactor score, the Article Influence score uses the structure of the entire citation network to evaluate the importance of each journal, does not consider self-citations and considers a time span of five years. It can be considered as a measure of the average influence of individual articles appearing in the same journal (Thomson Reuters, 2009), and it is normalised with respect to a mean of 1 (Bergstrom, 2007).

The Article Influence score is proportional to the Eigenfactor score, since it is calculated as follows:

$$\text{Article Influence score} = \frac{\text{Eigenfactor score}}{\text{Number of articles published}}$$

In order to illustrate the Eigenfactor metrics obtained by some of the journals belonging to the Science Citation Index of the 2007 Journal Citation Reports, let us consider the ten journals that obtained the highest Eigenfactor score (Table 7.16). The highest Eigenfactor score

Table 7.16 **Eigenfactor metrics of ten journals from the Science Citation Index (2007 JCR)**

Abbreviated journal title	Eigenfactor score	Article Influence score
Nature	1.83	16.99
P Natl Acad Sci USA	1.69	16.53
Science	1.69	16.53
J Biol Chem	1.53	2.28
Phys Rev Lett	1.26	3.21
J Am Chem Soc	0.95	2.63
Phys Rev B	0.78	1.25
Appl Phys Lett	0.71	1.49
New Engl J Med	0.69	17.86
Cell	0.67	18.18

obtained among the journals of the Science Citation Index is 1.83, corresponding to the journal *Nature*. This means that this journal receives an estimated 1.83 per cent of all the citation traffic in all of the journals listed by the Journal Citation Reports. Furthermore, this journal obtains an Article Influence score of 16.99, which means that its articles are, on average, almost 17 times as influential as the average article in the Journal Citation Reports (Bergstrom and West, 2008).

Note that those journals obtaining the highest Eigenfactor scores do not necessarily obtain the highest Article Influence scores. In addition, there may be differences when ranking journals according to Eigenfactor metrics or other impact measures. For instance, the journal *Nature*, which obtained the highest Eigenfactor score, is ranked tenth according to the traditional impact factor.

Finally, it is important to note some key considerations about the Eigenfactor metrics that have been identified by Bergstrom and West (2008). These authors stress that both the Eigenfactor score and the Article Influence score are aggregate measures of citation rates, i.e. they take into consideration all the papers published in a journal, aggregating its content. Therefore caution must be exercised when evaluating a single paper, since these measures do not focus on individual papers within journals. Bergstrom and West suggest that when assessing a single paper, it would be better to know the number of citations received by this paper rather than the Eigenfactor metrics or other impact indices of journals. In addition, they point out that citation data are not the only way to assess journals and other measures such as direct measures of readership and usage could be used to complement citation analysis.

Relationships between journals and areas

Citation analysis can be approached from the point of view of journals. In this case, the data considered are the references found in the publications of a given journal. Of course, the references found in the articles published in a journal will refer to different kinds of documents: other articles, books, proceedings, dissertations, etc.

Some research has been conducted regarding the different types of citations found in a journal. Biemans et al. (2007) analysed the references found in the *Journal of Product Innovation Management* over a twenty-year period and found that the percentage of non-journal reference sources had decreased significantly over time whereas the number of references to journals had increased. A possible explanation for this is that standard texts, manuals and practical books are less widely used nowadays than they were some years ago. Furthermore, the use of journals has been favoured by the increasing access to online journals. With respect to the increasing use of online resources, McDonald (2007) carried out a study of journal use in a science research university, the California Institute of Technology. He gathered the 1,521 journals owned by its library from 1997 to 2004 and studied whether they had been used in the printed version or through online access when this was available. The results showed that the use of a printed journal decreased significantly in terms of the consultations, photocopies or scans made when it was available in an online version. Thus, as expected, the availability of a journal in an online format results in less use of printed materials. Additionally, McDonald related the number of citations to the journal format: printed or online. Once again, online journals in this area were cited at a higher rate than were non-online journals, and citations increased after a journal became available online. Therefore, the online availability of a journal has a clear effect in terms of disseminating its content among the scientific community.

As regards citation analysis between journals, the procedure here is to count the number of citations from one journal to another. This analysis provides an idea about journals that are related to the same topic in a scientific area.

A basic and descriptive approach to citation between journals can be obtained by means of a rank of citations. Imagine that we want to analyse citations received by all the articles published by a journal over a given period of time. This will enable us to identify which journals are closely related to the journal we are studying, as those journals that cite it more frequently will share interests and contents.

The next example uses the journal *Health Education Research* to illustrate how to rank journals according to the number of citations they have made to another journal. Thus, in this case, we need to know which journals have most frequently cited articles from *Health Education Research*. Since this is a laborious task it is important to establish a time span for the analysis, which in this example is all articles published by this journal during 2005. As the calculations are based on citations received by these documents it is necessary to select documents for a period of time that is long enough to receive citations. In this case, the bibliographic search was conducted in the ISI Web of Science and was limited to articles. A total of 61 articles were found, and the total number of citations received by them was 421; this citation rate is sufficient for conducting the analysis. The aim is to rank those journals that have cited the articles published in *Health Education Research* during 2005; of course, citations made to this journal will be included in articles from other journals published since 2005 (i.e. 2005–9). The next step therefore is to make a list with all the sources that have cited these articles, as well as their frequency. In this case, a total of 234 sources, most of them journals, have cited the articles from *Health Education Research* published in 2005. Table 7.17 shows the journals that have most frequently cited the articles of this journal.

It can be seen that the journal which has given most citations to the journal studied is the journal itself. Thus self-citations seem to be important as well as those within authors. However, authors cite themselves to promote their own work, whereas journal self-citations serve to increase the impact factor.

By means of this procedure it has been possible to identify those journals that are most closely related to *Health Education Research*, as it receives a significant percentage of citations from this set of journals.

Another analysis based on journal citations involves ranking the most oft-cited journals in a given area of research. This can be illustrated by returning to the example based on articles found in the ISI Web of Science about *childhood epilepsy* between 2003 and 2008. A total of 94 articles were published in this field during this period of time and they have received 566 citations to date. These citations are distributed across 42 journals in this area. However, since we are interested in identifying the most oft-cited journal in this field, these journals must be ranked according to their citation frequency. Table 7.18 shows the 15 journals that have received the most citations during the established time period.

This information also enables us to analyse these data from another perspective: identifying the distribution of journals in zones according to

Table 7.17 **Ranking of the journals that more commonly cite the journal *Health Education Research* (2005 WoS)**

Source	Number of citations	%
Health Education Research	38	9.03
Patient Education and Counseling	10	2.38
BMC Public Health	9	2.14
Preventive Medicine	9	2.14
Annals of Behavioral Medicine	7	1.66
Journal of Health Communication	7	1.66
Prevention Science	7	1.66
Psychology & Health	7	1.66
Pediatric Exercise Science	6	1.43
Health & Place	5	1.19
International Journal of Behavioral Nutrition and Physical Activity	5	1.19
Transportation Research Part F – Traffic Psychology and Behavior	5	1.19
BMC Health Services Research	4	0.95
Cancer	4	0.95
Diabetes Educator	4	0.95
Health & Social Care in the Community	4	0.95
Health Psychology	4	0.95
Journal of Advanced Nursing	4	0.95
Social Science & Medicine	4	0.95

the number of times they have been cited. This will reveal a core of the most oft-cited journals. This idea of identifying a core set of journals is the same as in the case of Bradford's law regarding journal productivity. However, whereas in the section on journal productivity we sought to identify the most prolific journals, the objective here is to identify those journals that have received more citations by others. Nevertheless, the procedure used is the same as that described in Chapter 4.

The application of Bradford's law to citation analysis has been supported by Delwiche and Hall (2007). Here, we will apply Bradford's law to the ranking of journals in the field of childhood epilepsy (Table 7.18) and will include all those journals that have published in this field during the established period. Thus we need to list all the 42 journals

Table 7.18 Ranking of the most cited journals in the *childhood epilepsy* area (2003–9)

Journal	Times cited
Epilepsia	92
Seizure – European Journal of Epilepsy	58
Epilepsy & Behavior	55
Epilepsy Research	45
Journal of Paediatrics and Child Health	40
Pediatric Neurology	37
European Journal of Paediatric Neurology	33
Brain	25
Brain and Language	19
Journal of Physiology – London	18
Clinical Neurophysiology	16
Journal of Child Neurology	16
Epileptic Disorders	14
Lancet Neurology	12
Developmental Medicine and Child Neurology	11

that have published on this topic and rank them according to the number of citations they have received.

The total number of citations received by this set of journals is 566. The aim here is to divide these journals into three zones as this is the usual number of Bradford's law zones, each of which will contain approximately 33.3 per cent of citations (Table 7.19).

This table shows a core zone of the most oft-cited journals, comprising three journals that account for more than 30 per cent of citations received in this area. As we move away from the central zone, we need more journals to reach the same percentage of citations received. Journals found in the third zone will be the numerous journals that have received a small number of citations.

Relatedness index

Relationships between journals can be explored by means of the relatedness (*R*) index (Pudovkin and Garfield, 2002). *R* is determined for

Table 7.19 **Bradford's zones based on the references obtained in the *childhood epilepsy* field**

	Journal		Citations received	
Bradford zones	Number	%	Number	%
1	3	7.1	205	36.2
2	5	11.9	180	31.8
3	34	81.0	181	32.0
Total	42	100.0	566	100.0

a given period of time and takes into consideration two journals between which we want to know the degree of relatedness. Thus we need to know the number of citations in a given period of time from journal *A* to another journal *B*, the total number of citations from *A* to all journals and the total number of articles in *B*.

Thus the relatedness of *A* to *B* can be defined as:

$$R_{AB} = \frac{C_{AB}}{P_B \times CT_A}$$

where C_{AB} is the number of times that *A* cites *B*, P_B is be the number of articles published in journal *B* and CT_A is the total number of citations in *A* (to all journals). An example of this analysis is shown in Willett (2007), who applies the relatedness index to identify those journals with the closest relationship to the *Journal of Molecular Graphics and Modelling* (*JMGM*). This study is based on the 2007 Journal Citation Reports database, and the author shows the ten journals that obtain the highest *R* values and which are therefore the most closely related to the *JMGM* (Table 7.20).

As shown in this table, *R* has been calculated not only for the *JMGM* to other journals, but also from other journals to the *JMGM*. Related journals are ranked in decreasing order of the larger of the two possible values, as suggested by Pudovkin and Garfield (2002).

Thus we can identify which journals are closely related to a given target journal based on the citations that it makes and receives.

Finally, another way of studying the relationship between journals can be based on co-citation analysis. In the previous chapter we saw how co-citation analysis can identify the most oft-cited groups of authors. Here, the same analysis will be applied to identify the groups of journals that are most commonly co-cited.

Table 7.20 **Journals that are closely related to the *Journal of Molecular Graphics and Modelling***

	Relatedness (× 10^6) of	
Journal	***JMGM* to J**	**J to *JMGM***
Journal of Computer-Aided Molecular Design	250.35	256.16
Journal of Chemical Information and Modeling	62.95	186.84
Journal of Computational Chemistry	162.85	66.96
Structure	30.00	141.65
Proteins	55.99	116.33
Acta Crystallographica D	15.04	111.91
SAR and QSAR in Environmental Research	31.48	98.87
Journal of Molecular Modeling	22.36	96.27
Current Opinion in Structural Biology	84.66	41.70
Protein Science	26.56	79.73

Reprinted with permission from Willett (2007).

McCain (1990) was the first author to suggest using co-citation analysis with journals as the object of study. In this case, the data pairs included in the analysis will be the co-cited journals and the same procedures used for author co-citation analysis will be employed, i.e. cluster analysis or factor analysis. As before, it should be remembered that cluster analysis does not allow the same journal to be included in two different clusters as the latter are mutually exclusive. However, with factor analysis the same journal can load on more than one factor. This method will be useful for identifying interrelationships between specialities (Liu, 2005).

8

Important considerations

This chapter discusses the main limitations associated with bibliometric studies, which are largely related to the indices used to measure scientific output. As we have seen, most of the indices applied to assess quality and productivity are based on citation data, mainly those extracted from the ISI Web of Knowledge. Thus, although other sources of data and databases are available, most researcher and journal evaluations are based on the indices and citation counts provided by the ISI Web of Knowledge. The enormous influence that the interpretation of these indices has in practice has led many authors to seek to improve the calculation and use that is made of these analyses. As already mentioned and as we will see in this chapter, the use made of bibliometric indicators, basically citation analyses, is not always very accurate. Consequently, this section will discuss the main issues that have to be taken into consideration when applying bibliometric indicators, which are mainly focused on citation counts.

In this regard, Weingart (2005) warned of the very influential role of the ISI and argued that it had come to monopolise citation data analysis. However, and as van Raan (2005) states, the problem is not with the data producer but, rather, with those who misunderstand these measures. Although the use of new technologies has enabled the commercialisation of bibliometric data, it has also rapidly increased the non-expert use of bibliometric indicators (Adam, 2002).

Bibliometric analyses are often used as a measure of the quality of the work produced by an author, journal or department, etc. However, there is a strong support for the need to combine bibliometric indicators with peer-based evaluation procedures when assessing scientific activities (Moed and van Leeuwen, 1995; van Raan, 1996, 2000, 2005; van Raan and van Leeuwen, 2002). Bibliometric indicators are also used by libraries to decide which journals to purchase, and this will be based on their impact factor. Garfield (2006) states that, as a general rule, journals with high impact factors include the most prestigious ones. What is true is that publishers and editors celebrate an increase in their impact factor while a decrease can send them off into a huddle to figure out ways to boost their ranking (Simons, 2008).

In sum, bibliometric indicators need to be used carefully and correctly. As van Raan (2005) asserts, improper calculation and sloppy application of bibliometric indicators negatively influence the general appreciation of bibliometric methods by the scientific community.

Citation count

There is a general limitation when working with citations which is related to the way citations are gathered and counted. Working with the large databases used in bibliometric studies implies multiple data from different sources and these will obviously contain mistakes or incomplete entries. At times the source of the error will be difficult to identify, as one is dealing with large amounts of information about authors, institutions and countries, etc. Nevertheless, citation analyses are often used as a tool to evaluate scientific productivity, and this can have important consequences for the author, research group or university that is being evaluated. As such it is essential that every possible effort is made to ensure that the data are error free. This section considers the main mistakes that may be found in a typical database that might be used to analyse citations.

Moed (2002b) points out the importance of correcting the possible errors that might be found in citations and estimates that in a given study the percentage of citations gathered with incomplete information could be as high as 30 per cent of those included.

Faced with this evidence, van Raan (2005) summarises the main mistakes that may be found when working with citations. Firstly, he warns of the difficulty in attributing publications to specific organisations. Although we often try to find the affiliation of contributing authors there is no standard mechanism for including these affiliations in an article. While some authors indicate the name of their department, others only include the name of their university. Furthermore, other groups or institutes may prefer to mention their own name rather than the name of the university in which they work, and it can also be difficult to identify institutes within organisations. A clear example is the case of medical schools, hospitals or medical centres that are linked to a particular university, as it will be difficult to identify these different centres as belonging to the same university. Thus, if our aim is to assess the productivity of institutions, a raw citation analysis from the data gathered can lead to erroneous results. The question is therefore how to assess the productivity of a given university if we are not sure which departments or centres are part of it.

Another source of error is the different ways in which the name of the same university is written, a common mistake that was mentioned in Chapter 2 when analysing institutional productivity. As we have seen, the same university can be referred to in different ways depending on the article, the different languages spoken in a given area or simply the multiple ways in which it can be written (van Raan, 2005). Therefore, before carrying out a bibliometric study, it is necessary to ensure that author affiliations have been unified as an accurate identification of this possible bias in our database is the only way of avoiding a misinterpretation of results.

A further bias related to author citation analysis concerns the author name itself, since many authors will have the same surname, or even the same initials. This represents a serious problem when analysing author productivity, collaboration patterns or any other analysis based on author names (Moed, 2002b).

At times the same author may have his or her name written differently in articles, a common source of bias in this regard being the use of a middle name. In those areas in which middle names are not used, much confusion can arise between the surname and what, in fact, should have been a middle name. However, and as we will see in a following section, a system has been developed to overcome the problems related with author names.

Another possible source of bias is that related to the formal aspects of a journal, for example as a result of journal name changes or mergers (Moed et al., 1999).

Although a formal change should not influence the bibliometric indicators associated with a journal, it commonly happens in practice. A clear example is that shown previously in Chapter 7, where a change of title influenced the impact factor value. If we are considering the citations received by a journal in a given period of time, the title change will mean that, during a certain period, no citations will be counted for that journal. Consequently, the impact of the journal will fall dramatically after this formal change as it will appear to be receiving no citations.

Another formal aspect that can influence the citation count is the distribution and numbering of documents across volumes and numbers of a journal. In this regard, the journals most strongly affected in their citation count will be those with dual volume-numbering systems or combined volumes (Moed, 2002b).

All in all, these biases in reference counts could lead to a large proportion of citations in a field being lost (Moed, 2002b), and the only way to solve this problem is to be extremely accurate when working with citations and try to identify all possible mistakes in the citations gathered. Indeed, this is the only way to ensure that the analysis is really valid.

In addition to mistakes derived from the inherent characteristics of citations, other biases in the citation count should also be taken into consideration. For example, there is evidence regarding differences in the number of citations and their age across research fields. Snyder and Bonzi (1998) found significant differences in the average number of citations between disciplines, specifically that sciences and physical sciences had a higher frequency of citations. However, they reported that the pattern of self-citations did not differ between fields.

In summary, citation count can be influenced by many factors which may bias data, which must be carefully examined in order to minimise mistakes. A critical attitude must also be adopted as regards analyses based on citation counts, as interpretations derived from such analyses must be correctly targeted to the field and period under analysis.

Impact factor and other impact indices

The impact factor and other related indices, such as the immediacy index or the cited half-life, have become the subject of serious debate. Although these indices are widely used in practice it is necessary to consider their main limitations. Furthermore, given that research evaluation based on these indices has important implications for everyday practice, we must be aware of the particular factors that may influence their values. As the *impact factor* has also been used to assess scientists for hiring or promotion purposes, or in the case of research groups and institutions for funding (Fassoulaki et al., 2001), we have to be sure that the impact factor of a journal has been accurately obtained.

Some authors argue that the impact factor has an arbitrary nature, and they point out that a higher impact factor may not actually be a reflection of journal quality as it is supposed to be (Simons, 2008). Others emphasise the lack of rigour in selecting the number of publications involved in impact factor formulation (Haynes, 2007), or the absurd enslavement to impact factor that exists at present (Green, 2008). Consequently, although there are many criticisms of impact indices, most of them are centred on the journal impact factor, as this is the most widely used and the one that has the most consequences. This section will identify the main sources of bias that can influence these indices.

The first form of bias, one that has been widely debated, is the difference in impact factor values between *research fields*. It should be borne in mind, as van Raan (2005) points out, that journal articles are not the main vehicle for scientific knowledge in all fields. Thus journal articles will have neither the same meaning nor the same value across fields. Consequently, the impact factor may vary enormously between journals belonging to different fields or between different subject categories. Calculating the impact factor may therefore be valuable in terms of identifying journals with the highest impact in a given research field, but not for comparing journals between fields. In fact, the first formulation and use of the impact factor was limited to a single field, compiling references from a single key journal (Allen, 1929; Gross and Gross, 1927). Subsequently, impact factor analysis was extended to other sources, and included more journals and even more than one discipline (Brown, 1956; Gregory, 1937; Martyn and Gilchrist, 1968; McNeely and Crosno, 1930). However, these authors always presented scores separately for each research field. Thus it seems that in the early years of impact factor application authors were sensitive to the differences between fields as they produced field-specific listings. There was no need for cross-field comparability, as the purpose of the technique was to identify relevant journals for libraries to have adequate journal coverage of a series of different fields (Archambault and Larivière, 2009). Garfield (2006) has also pointed out the need to adjust impact indices to different research fields. However, not only the impact factor but also other indices such as the half-life can present differences between fields. Furthermore, this author agrees with the idea that the two-year target window used by the impact factor may, in some fields, fail to provide as suitable a period as would a five- or ten-year window. Nevertheless, and as pointed out previously, he also supports the idea that when studying journals by category, the rankings based on 1-, 7- or 15-year impact factors do not differ significantly (Garfield, 1998a, 1998b).

In sum, it is now generally accepted that impact factor can vary widely across fields, and as Althouse et al. (2009) summarise, there are three reasons why this might be so. Firstly, it can be due to differences in citation practices (Moed et al., 1985), as well as to differences in the delay between publication and subsequent citation (Marton, 1985; Moed et al., 1985). Finally, it could be a consequence of the differences in the proportions of citations made to literature indexed in the Journal Citation Reports (Hamilton, 1991; Vanclay, 2009).

Despite the evidence for the tremendous variation in the impact factor across fields, this is often not taken into consideration in research practice. As Archambault and Larivière (2009) state, abuse of the impact factor has been identified when it is applied across fields. One example of this is the fact that promotion decisions are sometimes based on raw impact factor values (Fuyono and Cyranoski, 2006). These authors identified cases in which bonuses were given to researchers depending on the cumulative one-year impact factor of the journals in which they publish. Considering that impact factor varies between fields, this measure would help researchers in some fields become wealthier (for instance, in the biomedical field, where there are higher citation rates and, therefore, higher impact factors). In contrast, researchers in other fields such as mathematics or social sciences would only obtain small bonuses, as impact factors in these disciplines are lower, even when publishing in the best journals of these fields.

Another traditional criticism of impact factor indices concerns the influence of *productivity* on their calculation. In this regard there is evidence for a linear relationship between journal productivity and impact factor (Rousseau and van Hooydonk, 1996). This correlation is based on the fact that those journals which publish more articles will obtain more citations and, therefore, higher impact factors. Related to this fact, some studies have demonstrated a relationship between the number of citations received and the impact factor (Cho et al., 1998). Furthermore, if we consider productivity as a whole applied to a given field, there is also evidence that the most productive fields (in terms of the number of journals) have the highest impact factors (Jemec, 2001).

Contrary to these criticisms, Garfield (2006) argues that productivity does not influence impact factor. He states that this widespread belief overlooks the fact that while more authors produce more citations (numerator), these must be shared by a larger number of cited articles (denominator). Moreover, he argues that the different type of documents taken into account in the numerator and denominator do not really affect the impact factor calculation. His reasoning is that although documents are not included in the denominator, the few citations they receive are counted in the numerator. Thus he considers that any resulting distortion is limited, and only a small group of journals will be affected.

This discussion of productivity leads into another question, namely how the impact factor can be increased by maximising the numerator of its formula (number of citations received) or minimising the denominator (number of items published by the journal). A further issue related to the impact factor formula is the debate about which items should be counted in its calculation. Let us consider these criticisms.

Firstly, it is necessary to return to the pattern of citations received by journals. Vanclay (2009) offers a humorous explanation of journal citation patterns and identifies what he calls hare and tortoise journals. The hares are those journals to which citations accrue quickly over a confined period of time, while tortoises are those journals to which citations accrue slowly over an extended period of time. Thus the number of citations received by these journals in the same period of time will be different. This fact is related to the period of time used to count the number of items published (target window). As we saw in Chapter 7, there is strong support for the idea that more than the previous two years should be considered when calculating an impact factor. Consequently, a five-year impact factor has recently been introduced in the Journal Citation Reports in order to overcome this bias which obviously favours hare journals. This bias also affects disciplines, as the two-year window used to estimate the journal impact factor generates an unequal sample for different journals, and thus the impact factor does not provide a comparable indication of impact across different disciplines.

However, debate continues as regards the number of citations and items that should be included in the impact factor analysis. Returning to the idea of productivity, Garfield himself (1996) recognised that absolute citation counts tend to favour older journals and those that publish more papers. Thus it seems that those journals which publish more papers will also receive more citations (increasing the numerator value) and, therefore, they will obtain higher impact factors.

By taking into consideration the items that will be included in the impact factor formula, journals could select the type of documents they publish in order to manipulate their impact factor (Simons, 2008). In the numerator of the impact factor formula, all the citations received by the journal are taken into account. However, the denominator only includes those documents identified as substantive articles in terms of citations, i.e. primary research articles and review articles. This means that other types of documents such as correspondence, letters, commentaries, perspectives, news stories, obituaries, editorials, interviews or tributes will not be included in the denominator (Garfield, 2006). However, the citations they receive are included in the numerator.

Some authors have reported biases in this respect. Simons (2008) warns that in many journals the number of reviews has increased dramatically, while Amin and Mabe (2007) note that the document type is also an important factor with respect to citation variation, even within the same subject area. This is because review articles are cited more often than are research articles, so the publication of reviews will increase a journal's impact factor. The fact that review journals often obtain higher impact factors has also been accepted by Thomson Reuters (1994). They attribute the higher number of citations received by review articles to their utility as a surrogate for earlier literature. The criteria established by the Journal Citation Reports to consider an article as a review are as follows: to belong to a review section of a journal or be coded as a review, or to contain more than 100 references. Consequently, review articles will increase the impact factor of a journal.

Simons (2008) also notes the growing interest in publishing commentary articles as the citations they receive will be counted in the numerator whereas this type of document will not be considered in the denominator. Consequently, a higher impact factor will be obtained when publishing a commentary article that receives citations.

In order to illustrate how the impact factor can change significantly due to the structure of a journal, Joseph and Hoey (1999) describe the case of *The Lancet*. The impact factor of this journal decreased significantly from 1996 to 1997 because during this period the number of items it published almost doubled, mainly as a result of the introduction of a Research Letters section in 1997. This section included short reports which were less likely to be cited. Thus the number of items counted in the denominator increased (as the articles from this new section were included as substantive scientific articles), whereas the number of citations received did not increase significantly, as research letters do not receive as many citations. As a result, the inclusion of this new section in *The Lancet* led to a reduction in its impact factor.

Another factor that can influence the count of the items to be included in the impact factor formula is the difficulty of classifying these items. Joseph and Hoey (1999) described another example of how the impact factor calculation can be misused, and they stressed the need for an accurate classification of the items published by journals. What they did was to compare the impact factor shown by the Journal Citation Reports for the *Canadian Medical Association Journal* with the impact factor they calculated manually. They found that the number of items considered by the JCR was higher than the number they considered for the hand-counting procedure, and as a result the official impact factor of this journal was lower. The discrepancy was due to the fact that the denominator used by the JCR to calculate the journal's impact factor was inflated by the inclusion of reports published in a given section as substantive articles. Luckily, this mistake was corrected and these documents were excluded from the later denominator count. This mistake was a consequence of the difficulty in classifying different types of items, and it is not the only reported case of this kind. For instance, in their study of the journal *Cortex*, Della Sala and Brooks (2008) found, during a given period, that this journal published a set of selected proceedings which were counted as source items, thus increasing the denominator value. This led to a decrease in the impact factor value over the following two years. Once again, it was necessary for the journal's editors to raise strong objections before these proceedings were excluded from the substantive articles count in the Journal Citation Reports.

Another source of criticism is the inclusion of self-citations in the impact factor count. The effect of self-citations on the bibliometric indicators has already been discussed with respect to both authors (Chapter 6) and journals (Chapter 7), and so it will only be briefly mentioned here. Self-citations of a journal are included in the citation count used in the impact factor formula, where it can inflate the numerator. Thomson Reuters (1994) estimate that approximately 13 per cent of citations received by a journal are self-citations. However, higher self-citation rates have been reported. For example, Fassoulaki et al. (2000) investigated how the frequency of journal self-citations affects the impact factor in journals related to *anaesthesia*, their own area of research. As expected, they found a correlation between the number of self-citations and the impact factor, but they also observed surprisingly high self-citation rates. When analysing the total number of references made by one of the journals studied over a period of two years, they discovered that 57 per cent of them were self-citations. Furthermore, their count of the number of self-citations received by these journals, and involved in the impact factor formula, revealed rates as high as 30 per cent. This clearly illustrates how self-citations can produce important increases in the impact factor. However, the key problem with self-citations is the way in which journals misuse them. As Archambault and Larivière (2009) point out, a journal can try to increase its impact factor by encouraging authors who wish to publish in it to include references to the journal itself in their articles. Furthermore, rather than being an isolated fact this appears to be common practice, and one might therefore ask what is actually being measured by the impact factor. If we really wish to assess the impact a journal has in the scientific community, authors should not be encouraged to make self-citations.

Another bias emerges when analysing self-citations regarding *research groups*. As van Raan states (2006a, 2008), group performance can influence the rate of self-citations received and the number of authors contributing to a paper (Della Sala and Brooks, 2008) is a factor that affects the number of self-citations received by the paper. Thus high-performance groups and multi-authored papers are more likely to be self-cited and will therefore receive a higher total number of citations. As before, these effects on the citation count will have repercussions for the calculation of bibliometric indicators.

In addition to the severe criticism of the impact factor as a measure, the application and interpretation of *cited half-life* has also been questioned. This topic is discussed in detail by Moed et al. (1998, 1999), who consider, specifically, that cited half-life insufficiently reflects the ageing characteristics of a journal. Their main criticism of this index is that it is not corrected to the size of citable articles of a journal over the years, so once again journal productivity comes into play. Other authors have demonstrated a relationship between journal productivity and cited half-life (de Queiroz and Lancaster, 1981; Wallace, 1985), and Moed et al. (1999) suggest adjusting cited half-life according to the number of publications of the journal. They calculated their own corrected cited half-life for a group of journals from the ISI Web of Knowledge and compared the results to those shown in the Journal Citation Reports. Specifically, rather than calculating the cited half-life according to the period of time that accounts for 50 per cent of citations, they did so according to the period of time in which the average impact of the journal was significantly reduced. Thus they related the journal impact factor to the cited half-life. In calculating this index they corrected the results by the difference in the number of documents published per year, obtaining what they called the *corrected cited half-life*. Their corrected index yielded values that were significantly different to those calculated in the Journal Citation Reports.

The *immediacy index* has also been shown to be related to journal productivity, and there is evidence for a relationship between impact indices (i.e. impact factor, cited half-life and immediacy index) and productivity (Ren and Rousseau, 2002).

The final source of bias to be discussed in this section is related to *publication delays*. This factor influences not only the impact factor, but also the immediacy index and cited-half life. Publication delay refers to the average delay in the publication of articles in a single scientific research field.

Yu et al. (2005) proposed the concept of the delay effect on literature citation and, based on simulated data, suggested that there is an inverse relationship between a field's average publication delay and the corresponding journal impact factor. The influence of publication delays on citation distribution has also been demonstrated by other authors (Garfield, 1999; Luwel and Moed, 1998; Marchi and Rocchi, 2001; Yu et al. 2006).

The repercussions of publication delay on bibliometric indicators are summarised by Yu et al. (2006), who showed that the longer the publication delay is in an area, the lower the impact factor and immediacy indices are for a given journal. The relationship between publication delays and the citation count used in these indices is clear. If manuscript refereeing or processing is delayed, then references to articles that are not within the two-year impact factor window will not be included in the count of these indices (Yu et al., 2005). Indeed, one would expect that articles published after a long delay will contain references that will then not be counted when calculating these indices. Thus there will be fewer citations over the last two years as well as within the same year of publication.

Another consequence of longer publication delay is that it yields a higher cited half-life index (Yu et al., 2006), as the average age of citations will increase.

To conclude, it is important to bear in mind the point made by Simons (2008), i.e. that above and beyond numerical methods for evaluating research quality, what really counts is the quality of a scientist's work wherever it is published. Consequently, impact factors should not be treated as direct measures of quality and they should always be used carefully (Amin and Mabe, 2007).

Possible solutions to citation count and impact factor indices

This section summarises the main solutions put forward to overcome the limitations associated with citation analysis, on which impact indices are based.

The first proposal concerns an available resource for avoiding the problems associated with author names. As we have seen, these problems are very difficult to resolve, and identifying authors when more than one share the same surname and initials, or when the same person finds his or her name written in different ways, is a laborious data-sorting task. A good solution is the resource introduced by the ISI Web of Knowledge known as *Researcher ID*. This is an online database in which researchers can be identified by a unique number. By logging in, authors obtain a researcher ID number that identifies them and they can then select their own published works, either directly from the ISI Web of Science or from other sources. Using this application an author can manage his or her publications, identifying those in which he or she has participated. Thus, despite there being more than one author with the same surname, or even in the case of misspelled names, the author can select the works in which he or she has actually collaborated. Once publications have been selected the collaboration network can be analysed, ranking those researchers with whom an author works. A ranking of the other researchers who have cited an author's work can also be obtained. Authors are therefore advised to obtain their researcher ID in order to avoid, as far as possible, the mistakes found in bibliometric studies due to author names.

In order to solve the problems associated with the application of the impact factor, some authors have proposed alternative indices. For example, Moed et al. (1999) developed a *normalised impact factor* to correct the differences in citation characteristics between sub-fields. They also proposed an *improved cited half-life* by developing a new index that takes into account other variables which influence this index, such as the speed of maturing and the decline in impact. Another proposal to normalise the impact factor between fields was made by Fassoulaki et al. (2002). These authors studied the impact factor of high-impact-factor journals belonging to seven medical research fields. As differences between fields will be found in terms of their impact factor, they proposed to normalise its value according to the field to which the journal belonged. The procedure used to calculate the *inter-field impact factor normalisation* was as follows: the mean of the upper quartile impact factor for each field was calculated, and the impact factor found for each journal was then divided by the mean of the upper quartile. By means of this simple ratio, differences between fields were reduced and more comparable values were obtained.

In a similar vein Althouse et al. (2009) applied a *weighted average impact factor* measure to improve the traditional assessment of the journal impact factor. This index is based on the relationship between the number of citations received and the number of articles published by a journal (productivity), and assumes that it is appropriate to assign higher weightings to journals that publish more articles (Egghe and Rousseau, 1996).

A further proposal is that of Vinkler (2004), who developed the *specific impact contribution* (SIC) to characterise the impact of a subset of articles or journals.

In addition, Jacso (2001) focused on the need to *improve the denominator of the impact factor* formula in order to compensate for the disadvantage it implies for less productive journals.

Finally, Habibzadeh and Yadollahie (2008) applied the *weighted journal impact factor* that considers the prestige of the citing journal. These authors support the idea that one of the limitations of the impact factor is that equal weight is given to all the citations received by a journal, without considering the prestige of the citing journal (Kochen, 1974; Pinski and Narin, 1976). They therefore propose a weighted impact factor that provides a more accurate idea about the quality and importance of a journal; specifically, it gives greater weight to citations coming from a prestigious journal than to citations made by a low-profile journal. The procedure used by these authors to weight citations is based on the comparison of the prestige of the citing and cited journal. In other words, they compare the impact factor received by the cited and citing journal in the previous year in order to know which is the most prestigious, and citations are then weighted. The value obtained from the comparison of the two journals is then normalised, obtaining weight ranges from 0.1 to 10. A prestigious journal receiving a citation from a low-prestige journal will obtain a weight of 0.1. In contrast, a low-prestige journal receiving a citation from a prestigious one will have the maximum weight, 10. As the authors point out, we should not compare values obtained by the traditional impact factor procedure and the proposed weighted impact factor, as the first emphasises the number of citations received while the second takes into account the prestige of the citing journal with respect to the cited journal. Once again, these bibliometric indicators have to be interpreted separately, as they focus on different aspects of the journal impact factor.

Hirsch index

As with the other indices presented, the h-index has inherent characteristics that can bias the values it yields. This section focuses on the application of the h-index to assess an individual's output, although the conclusions can be generalised to other levels of productivity. The aim of this section is merely to highlight several components of this index that should be taken into consideration when applying it.

It is worth mentioning the caveats that Hirsch (2005) himself pointed out regarding the application of his index. Firstly, it is important to bear in mind, as mentioned previously and supported by other authors (Costas and Bordons, 2007; Martin, 1996; van Raan, 2006b), that a researcher's evaluation should not be based on a single index. Indeed, it is always advisable to consider a range of indices when evaluating an individual, as decisions based on this evaluation can have important repercussions, such as the granting or refusing of tenure.

Regarding the potential bias in the h-index, let us begin with the *general interpretation* of this index. An author with a high h-index is considered to have published a high number of papers that have had a great impact. However, there are several variables that can directly influence the h-index value. For instance, there is evidence of a positive correlation between the h-index and the total number of papers published by an author or the total number of citations received (Costas and Bordons, 2007; Cronin and Meho, 2006; Hirsch, 2005; Jin et al., 2007; Kelly and Jennions, 2006).

In addition, a closer look at the papers on which the index is based provides a more accurate idea about the author's productivity and impact. As Hirsch (2005) points out, an author may have a relatively low h-index but this might not reflect his or her achievements, as he or she might have a few seminal papers with extraordinarily high citation counts (Hirsch, 2005) which may not be reflected in the h-index. In this respect, Costas and Bordons (2007) tested the influence of the absolute count of papers and citations on the h-index calculation and found that the h-index fails to identify those researchers who are very selective when choosing journals and who have intermediate levels of production but with a high international impact. Once again, if a researcher has published a few articles with high impact, his or her h-index will be small, as it will be bound by the small number of high-cited papers. Costas and Bordons therefore argue that the h-index undervalues the performance of researchers with an intermediate productivity level, but who may have a high impact and significant international visibility.

The relationship of the h-index to other indicators has also been demonstrated by other authors (Roediger, 2006; van Raan, 2006b). The fact is that the h-index obtained by two authors is not always comparable. Imagine the case of an author having $h = 12$. This mean that he or she has published twelve papers with at least twelve citations each. Now, if we compare this author with another who has $h = 6$, we will rapidly assert that the first has greater productivity and impact than the second. However, this may not be true. It is possible that the six papers belonging to the second author have had a great impact with numerous citations; for example, this small group of papers may have received more than 100 citations each. This should serve to remind us that caution must be exercised when asserting that one author has had a greater impact than another.

Another possible scenario is that two authors with the same h-index do not have the same performance. Imagine another case of two researchers, both of whom have an h-index of 12. Despite having the same number of articles with at least the same number of citations, the overall impact of their work may be different. For example, one of them may have twelve papers with 50 citations each, while the other has twelve papers with twelve citations each. Because of the inherent characteristics of the h-index it is always important to evaluate the productivity and impact of a researcher's work in depth, not only based on a single indicator.

Returning to the work of Costas and Bordons (2007), it is interesting to note the results they found regarding author impact classification. They analysed the impact and performance of researchers in their area of study (*natural resources*) in order to test whether the h-index overvalued or undervalued certain researchers depending on the absolute count of papers and citations. Their analyses were based on the comparison of the h-indices, divided into four categories or quartiles depending on what they termed the observed impact. This refers to the number of citations per document and the total number of citations received. They found that 16 per cent of the authors studied were undervalued by the h-index while 25 per cent were overvalued. Consequently, the h-index can be over- or undervalued depending on the publication pattern of a given researcher.

Another factor that can bias the value of the h-index is the *research field* to which it refers, and therefore the comparison of h-indices of researchers from different fields can lead to misinterpretation. As Hirsch (2005) states, there is considerable variation in the skewness of citation distributions even within the same sub-field. Thus those authors who work in highly topical areas will obtain a high number of citations and, as a result, will find it easier to achieve a high h-index. Hirsch tested these differences by analysing the h-index for the highly cited authors from different areas in the ISI Web of Knowledge. He found that h-indices were much higher in the life sciences and biological and medical sciences than in other fields such as physics.

Also related to the research field is the presence of *self-citations*, another source of bias that can influence the h-index. Zhivotovsky and Krutovsky (2008) warned that an excessive number of self-citations may be a common feature of some scientific fields and that excessive self-citations can inflate the h-index value. Regarding the research field, it should be remembered that the h-index must always be interpreted in comparison to other authors from the same research field.

Another issue related to specific research fields is the *number of authors* contributing to a paper, as some sub-fields tend to generate large-scale collaborations that will, in turn, produce higher h-values (Hirsch, 2005). As a result, authors with numerous articles in co-authorship may obtain higher h-indices. As we have seen, the number of self-citations will influence any of the indices based on citation counts, and given that authors tend to cite their own works, those papers written in collaboration with other authors will also receive more self-citations. These papers will therefore obtain higher absolute citation counts. Consequently, Hirsch proposes a normalisation of h-values to compare authors who show large differences between the co-authoring of papers.

In addition to the influence that self-citations have on multi-authored papers, special attention must be paid to whether or not they should be included in the citation count to calculate the h-index. Hirsh (2005) argues that self-citations should be excluded from any measure related to citations, including the h-index. However, he considers that despite increasing the h-index, the effect of self-citations on it is much smaller than that on the total citation count. This is because self-citations to an article with $< h$ citations are irrelevant, as they will be excluded from the analysis. Self-citations to papers obtaining many more citations than the h value will also be irrelevant. Nevertheless, some self-citations will obviously be taken into account and may increase the author's h-index. As we have seen, self-citations will not influence in the same way all the papers published by an author when calculating his or her h-index. Consequently, as Hirsch (2005) asserts, scientists who are intent on increasing their h-index can manipulate self-citations with the aim of increasing this index. This can be done by self-citing those papers with just $< h$ citations.

A simple example will suffice to show how self-citations can inflate the h-index. Let us consider the hypothetical case of an author having an h-index of 10. This means that he has ten papers with at least ten citations each, while his other published articles have fewer than ten citations. If this author cites those articles that already have ten or more citations this will have no effect on the h-index, as there will still be ten articles with ten or more citations. However, if the author cites those articles that are on the verge of being included in the h-index calculation, he can create a situation whereby he now has eleven articles with at least eleven citations each, thus obtaining $h = 11$.

For this reason, Zhivotovsky and Krutovsky (2008) reiterated the need to exclude self-citations from the citation count used in the h-index analysis, as well as from other impact indicators. Their results show that while modest self-citations do not greatly affect the h-index, excessive self-citations can inflate the h-index in the long term.

In summary, the h-index has been proposed as an accurate measure of individual performance, it being really useful when assessing individuals or journals. However, as with all measures based on citations, it has limitations due to the inherent characteristics of the count procedure. As has been repeatedly pointed out, the only way to ensure its correct use is to support any findings with other indicators and to exercise extreme care when comparing researchers according to their h-index.

Possible solutions to the h-index

In recent years a number of suggestions for improving the h-index have appeared. These are the subject of this section.

The first proposal for improving the h-index was made by Hirsch (2005) himself, who suggested using the *m value*. As the maximum h-index that a researcher can obtain will be the total number of papers published by him or her, researchers with different lengths of career will not always be comparable. The *m* value is obtained by dividing the h-index by the period of time elapsed since the author's first publication, and as such it considers the length of the author's career. This index is used to compare individuals of different scientific ages and can be thought of as the speed with which a researcher's h-index increases (Kelly and Jennions, 2006).

The *m* value is easy to calculate once we know the length of an author's career, i.e. the period during which he or she has been publishing. Imagine that a given author has an h-index of 15, and that 15 years have elapsed since he or she first published a paper. In this case $m = 1$. Hirsch (2005) proposes a criterion to interpret *m* values, whereby $m = 1$ corresponds to a successful scientist, $m = 2$ characterises outstanding scientists and $m = 3$ refers to truly unique individuals.

However, as Hirsch (2005) points out, an author's first paper may not always be adequate to establish his or her scientific age for the calculation of *m*. This is because the first paper might not really correspond to the start of his or her career if it represents a relatively minor early contribution prior to the subsequently significant productivity.

Another measure developed as a result of the h-index is the *g-index* (Egghe, 2006a, 2006b), which is defined as the highest rank such that the cumulative sum of the number of citations received is larger than or equal to the square of this rank. Consequently, $h \leq g$. The limitations and advantages of the g-index with respect to the h-index will not be detailed here as they have already been discussed in Chapter 6.

Another index that has been proposed is the *h(2)-index* (Kosmulski, 2006), which gives more weight to highly cited papers. A researcher's h(2)-index is defined as the highest natural number such that his *h(2)* most cited papers each received at least $h(2)^2$ citations. This means that an author with an h(2)-index equal to 10 will have 10 papers with at least 100 citations each. The h(2)-index will always be lower than the h-index, as it only identifies highly cited papers which will be fewer in number. The main advantage of this index is that it reduces the problem of accuracy, i.e. the lack of sensitivity of the h-index to performance changes (Jin et al., 2007). Burrell (2009) analysed the relationship between the h-, g-, and h(2)-indices and found that all three are approximately directly proportional to the length of the scientist's career, and also that they are proportional to each other. The relationship between them is as follows: $g \geq h \geq h(2)$.

Jin proposed the *A- and R-indices* (Jin, 2006, 2007). The A-index achieves the same goal as the g-index, namely correcting for the fact that the original h-index does not take into account the exact number of citations of articles included in the h-core. The A-index is defined as the average number of citations received by the publications included in the Hirsch core (Jin et al., 2007). Thus $h \leq A$. The A-index is obtained by means of the same ranking of papers used to obtain the h-index and therefore faces the same problem of accuracy. As we will see in an example below, the A-index presents an important limitation in that it sometimes disadvantages scientists with a higher h-index. However, this problem is easily solved by the R-index, which offers a correction of the A-index. The h-, A- and R-indices are related through the following simple equation (Jin et al., 2007): $R = \sqrt{A \times h}$. Once again, $h \leq R$.

Since these theoretical definitions of the A- and R-indices are quite abstract, let us consider a hypothetical example of their application. Imagine that we have two scientists, X and Y, for whom we can obtain their h-, A- and R-indices. Let us suppose that the first scientist has published a total of 20 papers, while the second has published 10. Regarding the citations received, we will assume that each of the scientists has a big-hit paper with a high number of citations, while their remaining papers have obtained the same, limited number of citations. Thus scientist X has a big-hit paper with 16 citations while his remaining articles have received two citations each. On the other hand, scientist Y has a big-hit article which has received 14 citations while the rest of his papers have been cited once each. This information is summarised in Table 8.1.

With this information we can easily obtain the corresponding h-indices. For scientist X, $h = 2$ as he has two papers with at least two citations each. However, scientist Y has a lower h-index ($h = 1$), which means that he has one article with at least one citation. The A-index is given by the average number of citations received by the papers included in the Hirsch core. As

Table 8.1 **Comparison of h-, A- and R-indices**

	Scientist X	Scientist Y
Number of papers	20	10
Citations (highest cited paper)	16	14
Citations (rest of papers)	2	1
h-index	2	1
A-index	9	14
R-index	4.24	3.74

we have seen previously, the Hirsch core corresponds to all the papers ranked between rank 1 and rank h. In this case, the Hirsch core of scientist X will be formed by the two highly cited papers, since $h = 2$. The A-index for scientist X is therefore $A = (16 + 2)/2 = 9$. For scientist Y, the number of papers forming the Hirsch core is one, and this paper has received 14 citations. Consequently, the A-index will be calculated as: $A = 14/1 = 14$.

These results are shown in Table 8.1. Considering the number of papers and citations that each of these hypothetical scientists has, it is obvious that scientist X is more productive and has a greater impact than scientist Y, and this fact is reflected by his higher h-index. However, a different result is obtained when applying the A-index, since scientist Y obtains a significantly higher A-index than does scientist X ($A = 14$ versus $A = 9$). It seems, therefore, that the A-index has punished scientist X for having a higher h-index, and this is why the corrective R-index was developed. As we have seen, once we know the h- and A-indices it is easy to derive the R-index. For scientist X, $R_X = \sqrt{9 \times 2} = 4.24$, while for scientist Y, $R_Y = \sqrt{14 \times 1} = 3.74$. Thus comparison of these two researchers by means of their R-indices shows that scientist X has the best performance.

In relation to these indices, Jin (2007) proposed the *AR-index* as an adaptation of the R-index, the aim being to overcome the problem that the h-index may never decrease. The difference between the two indices is that while the R-index measures the intensity of Hirsch core citations, the AR-index also takes the age of publications into account. This author proposes using this index with respect to a given period of a researcher's career rather than taking the whole career into account as in the case of the h-index. In fact, he supports the idea of using both the h- and AR-indices as a meaningful scientometric indicator.

The main advantage of the AR-index is that by taking into account both the actual number of citations and the age of publications it can increase or decrease over time, whereas the h-index will remain fixed for the period selected and only increase during an author's career.

Finally, and more recently, Egghe and Rousseau (2008) proposed the *weighted h-index*, denoted as h_w. This index weights the h-index according to the impact of citations and is sensitive to performance changes. Performance changes refer to the fact that the h-index can never decrease and that it is only weakly sensitive to the number of citations received. The relationship between the h_w and the other indices shown in this section can be written as follows: $h \leq h_w < g$; $h_w < R$.

In sum, various indices have been proposed to complement the h-index and to obtain a more accurate measure of individual performance. By knowing the advantages and disadvantages of these indices, as well as taking into consideration the nature of a specific case, researchers will be better placed to choose the most suitable index for a given analysis.

9

Final considerations

This book has discussed the main applications of bibliometric studies and has illustrated the great interest in applying quantitative methods to assess research performance, whether of individual scientists, research groups, journals or other levels of production. The increasing costs of research and the scarce economic resources available make research assessment essential for policy-makers (Costas and Bordons, 2008). Consequently, many analyses and indices, most of them based on citation analysis, have been developed to achieve this objective.

Citation analyses are usually based on the data provided by the citation indices of the ISI Web of Knowledge, developed by Thomson Scientific. Indeed, most of the examples given in this book are based on the JCR Science Citation Index (SCI) and Social Sciences Citation Index (SSCI). However, it should be noted that this is only one of the perspectives that a bibliometric study can adopt. For many years the citation indices produced by Thomson Scientific were the only source to rely on when assessing researchers or journals and they thus enjoyed a unique and privileged position within bibliographic databases. However, there are now other services that offer citation-enhanced databases and the electronic availability of documents has also given rise to new opportunities in automatic indexing. Consequently, many discipline-oriented databases have been launched in recent years (Neuhaus and Daniel, 2008), although there are two new multidisciplinary databases that merit particular attention. Both of them appeared in 2004 and they have an important content coverage: *Scopus* and *Google Scholar*.

The present chapter offers a broader overview of databases that can be useful when carrying out a bibliographic search for a bibliometric study.

Beyond the Web of Science: Scopus and Google Scholar

The Thomson Scientific databases have held an outstanding position among bibliographic databases. Their origins are attributed to E. Garfield (1955), who proposed an innovative indexation system based on a multidisciplinary approach, one which provided unique possibilities of studying multi- or interdisciplinary research activities. In 1963, Garfield proposed the Science Citation Index, which presently forms part of the Journal Citation Reports (JCR) published by Thomson Scientific, formerly the Institute for Scientific Information (ISI). Throughout this book we have become familiar with a number of concepts such as the ISI Web of Knowledge, which is a research platform for finding, analysing and sharing information in both the sciences and social sciences. The *Web of Science* provides access to the leading citation databases, covering nearly 9,300 of the highest impact journals worldwide. Finally, the Journal Citation Reports (Thomson Reuters) use citation data from over 7,500 scholarly and technical journals, including all areas of science, technology and social sciences. Certain citation information is also available regarding journals that are not covered, as well as those that are. The Journal Citation Reports have two editions: the Science Edition and the Social Sciences Edition, the first covering over 6,400 journals and the latter over 1,800. Overall, they provide cited and citing journal statistics from 1997 until the present moment, and once a year they publish journal statistics for the previous year of coverage.

However, this is not the only indexation system available to search or analyse bibliographies. Since 2004, researchers have also been able to use *Scopus* (published by Elsevier). This is the largest abstract and citation database, since it covers over 16,000 peer-review journals as well as conference proceedings, trade publications, book series and patents. The subject area coverage is distributed as follows: over 3,400 titles from the life sciences, over 5,300 in health sciences, more than 5,500 from the physical sciences and over 2,800 regarding the social sciences. The main strength of this database is that it offers the broadest coverage available for scientific, technical, medical and social sciences. Moreover, its coverage is worldwide, gathering content from Europe, Latin America and the Asia Pacific region. It also includes references from 1996, as well as additional cited references that are not covered by the database. In addition, most records as far back as 1823 have an abstract. Finally, it includes in-press articles from certain publishing groups so these are then available in Scopus prior to their official publication date.

Another source that has also been available since 2004 is *Google Scholar*, a service provided by Google. This offers a simple way of conducting a broad search for scholarly literature and can search across many disciplines and sources: peer-reviewed papers, theses, books, abstracts and articles, from academic publishers, professional societies, preprint repositories, universities and other scholarly organisations. When searching, Google Scholar sorts articles by weighting the full text of each article, the author, the publication in which the article appears and how often the piece has been cited in other scholarly literature. Thus the most relevant results will always appear first. The main strengths of Google Scholar are its free access (unlike the Web of Science and Scopus) and the availability of full texts. Indeed, Google Scholar provides a free view of abstracts to articles that sometimes require a fee subscription to view the full text. However, the data sometimes come from preprint servers or personal websites, and thus in many cases the full text is freely available to all users (Bar-Ilan, 2008). Google Scholar has agreements with publishers to allow searches of the full text of their articles (Butler, 2004). However, Pauly and Stergiou (2005) emphasise that Elsevier, the publisher that owns Scopus, has refused Google Scholar permission to search its texts. For publishers it is very easy to include publications in Google Scholar, as all they need do is send the abstract of a publication. Although the tool does gather the number of citations received by articles, it fails to include some information, for example the number of documents or journals it includes or its time coverage. Finally, it is important to bear in mind that Google Scholar is still a beta version, and consequently a number of improvements will be made before launching its final version.

Current position of multidisciplinary databases

The rigour and accuracy of a bibliometric study will depend, primarily, on the bibliographic search we carry out. If the aim is to analyse scientific productivity in an area of research we must be sure that this area is well represented in the database on which the bibliographic search is based.

Basically, two main aspects should be taken into consideration when selecting a database: its *coverage* and the *accuracy of its data*. Coverage relates to the extent to which the sources processed by a database cover the written scholarly literature in a field. As such, coverage must not be biased towards particular countries, languages or publishers (Neuhaus and Daniel, 2008). As regards the accuracy of data, this refers to the absence of inconsistencies and erroneous spellings of author names, or a lack of standardisation with respect to journal titles and affiliations (Neuhaus and Daniel, 2008).

When obtaining the raw data of a bibliographic search it is necessary to carry out a content analysis of the documents found. This will lead us to identify publications that may not be related to our field of study, or mistakes in the variables that we wish to analyse: authors, institutions, journals, years, number of citations, etc. Basically, the aim is to minimise any mistakes related to the accuracy of the data provided by the database, as well as to its coverage. Although a manual revision of data should be carried out prior to the final analysis, conducting the search in a database known to include good coverage and few spelling mistakes makes our task easier.

Now we will look at the reception which the Web of Science, Scopus and Google Scholar have received in the scientific community, as well as the differences found when conducting an analysis via these three sources. Of course, many databases specialised in particular fields are available, but this section focuses on these three databases due to their multidisciplinary nature and current relevance.

Let us begin with the general impressions about these databases. The Web of Science has the advantage of being based on the first multidisciplinary citation indexation and, to date, it has enjoyed a unique position among bibliographic search tools. The indices provided by the ISI Web of Knowledge regarding journal impact or citations are the mostly widely used in the assessment of scientific productivity, and this has led to the Web of Science having a privileged situation. However, its status has also been criticised, since impact indices have become the subject of a wide debate. Furthermore, some authors have criticised the enormous influence that the ISI Web of Knowledge now has in terms of scientist and journal evaluation (Green, 2008; Haynes, 2007; Simons, 2008).

The Scopus database has been well received by the scientific community, as it is considered to offer wide coverage and be easy to use (Burnham, 2006; Falagas et al., 2008; Jacso, 2008; LaGuardia, 2005). However, it has not been exempt from criticism as it seems that Scopus is not always successful when differentiating between authors with the same name (Bar-Ilan, 2008).

Finally, Google Scholar has received mixed reviews within the scientific community. As Bar-Ilan (2008) notes, while some users have responded positively (Giles, 2005), it has been less enthusiastically received by librarians, based on the small number of library sites linked to Google Scholar (Mullen and Hartman, 2006). As we will see, this database has come in for strong criticism, although some authors have argued in favour of its usefulness and potential (Gardner and Eng, 2005; Pauly and Stergiou, 2005; Rahm and Thor, 2005). These authors consider Google Scholar to be a competent database, comparable to the Web of Science, since it is free and covers proceedings that in some disciplines represent an important source of data, despite being partially indexed in the Web of Science. They also point out that it is still at the beta testing stage and thus has the potential to improve significantly over the next few years.

As we saw previously a key aspect to consider when evaluating and selecting a database is its coverage. In a general comparison of these three databases, Jacso (2005) found inconsistencies in all of them. Analysing the number of citations received by the same paper according to each of the three databases, he found they yielded three different citation counts. Moreover, only 40 per cent of these citations were the same in the three searches. However, this is not the only evidence about differences in their coverage. Bar-Ilan et al. (2007) found that citation counts were more similar between the Web of Science and Scopus than between Google Scholar and the other two databases, arguing that this reflects the different content coverage of Google Scholar in comparison with the Web of Science and Scopus. Other studies have also found that Google Scholar has a different coverage to that of the Web of Science and Scopus. As shown by Bauer and Bakkalbasi (2005) and Bar-Ilan (2008), there is evidence that Google Scholar shows considerably higher citation counts than the other two databases. Jacso (2006) attributes the inconsistencies found in Google Scholar citation counts to the fact that it also indexes non-scholarly sources. These strong criticisms contrast with the good agreement in citation counts found by Belew (2005) when comparing the Web of Science and Google Scholar.

However, the fact that Google Scholar has different coverage to the other two databases is not necessarily negative. As Pauly and Stergiou (2005) state, this database has the advantage of indexing abstracts from many major publishers, gathering research publications such as journal articles, books, preprints and technical reports.

Specifically in the social sciences, the Web of Science has been criticised for its insufficient coverage of this field (Norris and Oppenheim, 2007). This lack of coverage could be due to the fact that many citations in social sciences are made to books and monographs

(Hicks, 2004), as well as to articles published in non-English journals (Nederhof, 2006), which are not covered in the Web of Science. Norris and Oppenheim (2007) suggest using Scopus as an alternative to the Web of Science for evaluating research impact in the social sciences.

The coverage of a database will directly influence the citation counts and, consequently, the indices derived from them. One of the indices used to assess individual productivity is the h-index, which is based on the number of citations that an author has received. In this regard, Bar-Ilan (2008) analysed whether the h-index differs when taking into consideration data from the Web of Science, Scopus or Google Scholar. She found that, in general, there were no significant differences in the h-index calculation based on the Web of Science and Scopus. In contrast, the differences between Google Scholar and the other two databases were much more considerable. Once again, it seems that the content coverage of Google Scholar could differ from that of the other two. Bar-Ilan also found that even when authors obtained comparable h-indices with the data from the Web of Science, Scopus and Google Scholar, the respective citation counts could differ.

Related to the coverage of citation counts is the time span for the cited references. The Scopus database only covers citations from 1996, a fact that has been considered a limitation in contrast to its wide coverage of indexed documents (Jacso, 2008; Norris and Oppenheim, 2007).

The second main characteristic to be considered when selecting a database is the accuracy of its data. Although it is impossible to eliminate all erroneous data from databases, we should select a database that minimises mistakes in its indexed documents and their citations. As we saw previously, mistakes in references and citations will influence any subsequent bibliometric analysis, whether this be related to productivity or quality.

Jacso (2008) regards as positive the promising system used by Scopus to minimise errors related to authors and institutions. The database that seems to receive the most criticisms about the accuracy of its data is Google Scholar. Some studies have found inconsistencies regarding author names, as well as in the identification of publication dates (Jacso, 2005, 2006; Bar-Ilan, 2008). Google Scholar has also been criticised regarding its accuracy and for the fact that it displays items in relation to the number of user visits rather than in relation to other quality indices (Falagas et al., 2008).

In summary, as these three databases have different coverage and provide different and complementary resources, the decision as to which one to use may well depend on the aim of our study. As Bakkalbasi et al. (2006)

state, it is sometimes difficult to identify the best database for our specific case and, consequently, the complementary use of more than one search tool may be advisable (Levine-Clark and Gil, 2009; Meho and Yang, 2007).

Final thoughts

A key objective of this book has been to illustrate the main analysis used in bibliometrics. Since bibliometric studies have a wide range of applications, they can be adapted to each specific case. Therefore readers can select from among the analyses presented herein those which they consider to be most suitable for their specific bibliometric study.

Given that bibliometric indicators are widely used to assess (mainly) authors and journals, it is essential that steps are taken to ensure they are applied accurately. Furthermore, bibliometric indicators should not be used in isolation as a measure of performance. At all events, the present book's description and discussion of the main analyses used in bibliometric studies should help researchers to apply and interpret them from a critical perspective.

References

Adam, D. (2002) 'The counting house', *Nature*, 415: 726–9.

Aksnes, D.W. (2003) 'A macro study of self-citations', *Scientometrics*, 2: 235–46.

Allen, E.S. (1929) 'Periodicals for mathematicians', *Science*, 70(1825): 592–4.

Almind, P. and Ingwersen, P. (1997) 'Informetric analyses on the World Wide Web: methodological approaches to "Webometrics"', *Journal of Documentation*, 53: 404–26.

Althouse, B.M., West, J.D., Bergstrom, C.T. and Bergstrom, T. (2009) 'Differences in impact factor across fields and over time', *Journal of the American Society for Information Science and Technology*, 60(1): 27–34.

Amin, M. and Mabe, M. (2007) 'Impact factors: use and abuse', *Perspectives in Publishing Elsevier*, 1: 1–6.

Andrés, A., Gómez, J. and Saldaña, C. (2007) 'The transtheoretical model and obesity: a bibliometric study', *Scientometrics*, 73(3): 289–301.

Archambault, E. and Larivière, V. (2009) 'History of the journal impact factor: contingencies and consequences', *Scientometrics*, 79(3), 1–15.

Bakkalbasi, N., Bauer, K., Glover, J. and Wang, L. (2006) 'Three options for citation tracking: Google Scholar, Scopus and Web of Science', *Biomedical Digital Libraries*, 3: 7.

Ball, P. (2005) 'Index aims for fair ranking of scientists', *Nature*, 436: 900.

Banks, M.G. (2006) 'An extension of the Hirsch index: indexing scientific topics and compounds', *Scientometrics*, 69(1): 161–8.

Barabási, A.L., Jeong, H., Néda, Z., Ravasz, E., Schubert, A. and Vicsek, T. (2002) 'Evolution of the social network of scientific collaborations', *Physica A: Statistical Mechanics and Its Application*, 311(3–4): 590–614.

Bar-Ilan, J. (2008) 'Which h-index? A comparison of WoS, Scopus and Google Scholar', *Scientometrics*, 74(2): 257–71.

Bar-Ilan, J., Levene, M. and Lin, A. (2007) 'Some measures for comparing citation databases', *Journal of Informetrics*, 1(1): 26–34.

Barrios, M., Borrego, A., Vilaginés, A., Ollé, C. and Somoza, M. (2008) 'A bibliometric study of psychological research of tourism', *Scientometrics*, 77(3): 453–67.

Bauer, K. and Bakkalbasi, N. (2005) 'An examination of citation counts in a new scholarly communication environment', *D-Lib Magazine*, 11(9). Online at: *http://www.dlib.org/dlib/september05/bauer/o9bauer.html.*

Beaver, D.D. and Rosen, R. (1978) 'Studies in scientific collaboration 1', *Scientometrics*, 1(1): 65–84.

Beaver, D.D. and Rosen, R. (1979a) 'Studies in scientific collaboration 2', *Scientometrics*, 1(2): 133–49.

Beaver, D.D. and Rosen, R. (1979b) 'Studies in scientific collaboration 3', *Scientometrics*, 1(3): 231–45.

Belew, R.K. (2005) 'Scientific impact quality and quantity: analysis of two sources of bibliographic data'. Online at: *http://arxiv.org/abs/cs.IR/0504036* (accessed 19 April 2009).

Bergstrom, C.T. (2007) 'Eigenfactor: measuring the value and prestige of scholarly journals', *C&RL News*, 68(5).

Bergstrom, C.T. (2009) 'Eigenfactor Project. Ranking and mapping scientific knowledge'. Online at: *http://eigenfactor.org.*

Bergstrom, C.T. and West, J.D. (2008) 'Assessing citations with the Eigenfactor Metrics', *Neurology*, 71: 1850–1.

Bergstrom, C.T., West, J.D. and Wiseman, M.A. (2008) 'The Eigenfactor metrics', *Journal of Neuroscience*, 28(45): 11433–4.

Biemans, W., Griffin, A. and Moenaert, R. (2007) 'Twenty years of the *Journal of Product Innovation Management*: history, participants, and knowledge stock and flows', *Journal of Product Innovation Management*, 24: 193–213.

Biglu, M.H. (2008) 'The influence of references per paper in the SCI to impact factors and the Matthew effect', *Scientometrics*, 74(3): 453–70.

Björneborn, L. and Ingwersen, P. (2001) 'Perspectives of webometrics', *Scientometrics*, 50(1): 65–82.

Bookstein, A. (1976) 'The bibliometric distribution', *Library Quarterly*, 46(4): 416–23.

Borgman, C.L. (1990) *Scholarly Communication and Bibliometrics.* Newbury Park, CA: Sage.

Boyce, B.R. and Pollens, J.S. (1982) 'Citation-based impact measures and Bradfordian criteria', *Collection Engineering*, 4: 29.

Bradford, S.C. (1934) 'Sources of information on specific subjects', *Engineering*, 137: 85–6.

Braun, T., Glänzel, W. and Schubert, A. (2006) 'A Hirsch-type index for journals', *Scientometrics*, 69(1): 169–73.

Brodman, E. (1944) 'Choosing physiology journals', *Bulletin of the Medical Library Association*, 32: 479–88.

Brookes, B.C. (1968) 'The derivation and application of the Bradford-Zipf distribution', *Journal of Documentation*, 24(4): 247–65.

Brookes, B.C. (1969) 'Bradford's law and the bibliography of science', *Nature*, 224: 953–6.

Brown, C.H. (1956) *Scientific Serials: Characteristics and Lists of Most Cited Publications in Mathematics, Physics, Chemistry, Physiology, Botany, Zoology, and Entomology*, ACRL Monograph No. 16. Chicago: Association of College and Research Libraries.

Burnham, J.F. (2006) 'Scopus database: a review', *Biomedical Digital Libraries*, 3(1). Online at: *http://www.bio-digilib.com/content/3/1/1*.

Burrell, Q.L. (2009) 'On Hirsch's h, Egghe's g and Kosmulski's h(2)', *Scientometrics*, 79(1): 79–91.

Butler, D. (2004) 'Science searches shift up a gear as Google starts Scholar engine', *Nature*, 432: 423.

Casey, D.L. and McMillan, G.S. (2008) 'Identifying the "invisible colleges" of the *Industrial and Labor Relations Review*: a bibliometric approach', *Industrial and Labor Relations Review*, 62(1): 126–32.

Chen, Y.S. and Leimkuhler, F.F. (1986) 'A relationship between Lotka's law, Bradford's law and Zipf's law', *Journal of the American Society for Information Science*, 37(5): 307–14.

Chiu, W.T. and Ho, Y.S. (2007) 'Bibliometric analysis of tsunami research', *Scientometrics*, 73(1): 3–17.

Cho, J.E., Cho, C.T. and Belmont, J.M. (1998) 'Learning to assess the value of infectious disease journals', *Journal of Microbiology, Immunology and Infection*, 31: 1–4.

Costas, R. and Bordons, M. (2007) 'The h-index: advantages, limitations and its relation with other bibliometric indicators at the micro level', *Journal of Informetrics*, 1: 193–203.

Costas, R. and Bordons, M. (2008) 'Is the g-index better than the h-index? An exploratory study at the individual level', *Scientometrics*, 77(2): 267–88.

Crane, D. (1972) *Invisible Colleges. Diffusion of Knowledge in Scientific Communities*. Chicago and London: University of Chicago Press.

Cronin, B. (1984) *The Citation Process: The Role and Significance of Citations in Scientific Communication*. London: Taylor Graham.

Cronin, B. and Meho, L. (2006) 'Using the h-index to rank influential information scientists', *Journal of the American Society for Information Science*, 57: 1275–8.

Davarpanah, M.R. and Aslekia, S. (2008) 'A scientometric analysis of international LIS journals: productivity and characteristics', *Scientometrics*, 77(1): 21–39.

de Candolle, A. (1873) *Histoire des sciences et des savants depuis deux siècles*. Geneva and Basle: H. Georg.

de Queiroz, G.G. and Lancaster, F.W. (1981) 'Growth, dispersion and obsolescence of the literature – a case study from thermoluminescent dosimetry', *Journal of Research Communication Studies*, 2: 203–17.

Della Sala, S. and Brooks, J. (2008) 'Multi-authors' self-citation: a further impact factor bias?', *Cortex*, 44: 1139–45.

Delwiche, F.A. and Hall, E.F. (2007) 'Mapping the literature of athletic training', *Journal of the Medical Library Association*, 95(2): 195–201.

Egghe, L. (1986) 'The dual of Bradford's law', *Journal of the American Society for Information Science*, 37(4): 246–55.

Egghe, L. (1990a) 'A note on different Bradford multipliers', *Journal of the American Society for Information Science*, 41(3): 204–9.

Egghe, L. (1990b) 'Applications of the theory of Bradford's law to the calculation of Leimkuhler's law and to the completion of bibliographies', *Journal of the American Society for Information Science*, 41(7): 469–92.

Egghe, L. (1991) 'The exact place of Zip's and Pareto's law amongst the classical informetric laws', *Scientometrics*, 20(1): 93–106.

Egghe, L. (2005) 'Zipfian and Lotkaian continous concentration theory', *Journal of the American Society for Information Science and Technology*, 56(9): 935–45.

Egghe, L. (2006a) 'How to improve the h-index', *The Scientist*, 20(3): 14.

Egghe, L. (2006b) 'An improvement of the H-index: the G-index', *ISSI Newsletter*, 2(1): 121–9.

Egghe, L. (2006c) 'Theory and practice of the g-index', *Scientometrics*, 69(1): 131–52.

Egghe, L. (2007) 'Item-time-dependent Lotkanian informetrics and applications to the calculation of the time-dependent h-index and g-index', *Mathematical and Computer Modelling*, 45: 864–72.

Egghe, L. (2008a) 'Modelling successive h-indices', *Scientometrics*, 77(3): 377–87.

Egghe, L. (2008b) 'Mathematical theory of the h- and g-index in case of fractional counting of authorship', *Journal of the American Society for Information Science and Technology*, 59(10): 1608–16.

Egghe, L. and Rousseau, R. (1990) *Introduction to Informetrics: Quantitative Methods in Library, Documentation and Information Science*. Amsterdam: Elsevier.

Egghe, L. and Rousseau, R. (1996) 'Average and global impact of a set of journals', *Scientometrics*, 36(1): 97–107.

Egghe, L. and Rousseau, R. (2008) 'An h-index weighted by citation impact', *Information Processing and Management*, 44: 770–80.

Fairthorne, R.A. (1969) 'Empirical hyperbolic distributions (Bradford-Zipf-Mandelbrot) for bibliometric description and prediction', *Journal of Documentation*, 25(4): 319–43.

Falagas, M.E., Pitsouni, E.I., Malietzis, G.A. and Pappas, G. (2008) 'Comparison of PubMed, Scopus, Web of Science, and Google Scholar: strengths and weaknesses', *FASEB Journal*, 22: 338–42.

Fassoulaki, A., Papilas, K., Paraskeva, A. and Patris, K. (2002) 'Impact factor bias and proposed adjustments for its determination', *Acta Anaesthesiologica Scandinavica*, 46: 902–5.

Fassoulaki, A., Paraskeva, A., Papilas, K. and Karabinis, G. (2000) 'Self-citations in six anaesthesia journals and their significance in determining the impact factor', *British Journal of Anaesthesia*, 84(2): 266–9.

Fassoulaki, A., Sarantopoulos, C., Papilas, K., Patris, K. and Melemeni, A. (2001) 'Academic anesthesiologists' views on the importance of the impact factor of scientific journals. A North American and European survey', *Canadian Journal of Anaesthesiology*, 48: 953–7.

Fowler, J.H. and Aksnes, D.W. (2007) 'Does self-citation pay?', *Scientometrics*, 72(3): 427–37.

Frame, J.D. (1977) 'Mainstream research in Latin America and the Caribbean', *Interciencia*, 2: 143–8.

Freeman, L. (1979) 'Centrality in social networks: I. Conceptual clarifications', *Social Networks*, 1: 215–39.

Fuyono, I. and Cyranoski, D. (2006) 'Cash for papers: putting a premium on publication', *Nature*, 441(7095): 792.

Gardner, S. and Eng, S. (2005) 'Gaga over Google? Scholar in the Social Sciences', *Library Hi-Tech News*, 22(8): 42–5.

Garfield, E. (1955) 'Citation indexes for science – new dimension in documentation through association ideas', *Science*, 122(3159): 108–11.

Garfield, E. (1963) 'Science Citation Index', *Science Citation Index 1961*, volume 1, pp. v–xvi.

Garfield, E. (1972) 'Citation analysis as a tool in journal evaluation', *Science*, 178: 471–9.

Garfield, E. (1996) 'How can impact factors be improved?', *British Medical Journal*, 313: 411–13.

Garfield, E. (1998a) 'Long-term vs. short-term journal impact: does it matter?', *Scientist*, 12(3): 10–12.

Garfield, E. (1998b) 'Long-term vs. short-term journal impact (part II)', *Scientist*, 12(4): 12–13.

Garfield, E. (1999) 'Journal impact factor: a brief review', *Canadian Medical Association Journal*, 161(8): 979–80.

Garfield, E. (2006) 'The history and meaning of the journal impact factor', *Journal of the American Medical Association*, 295(1): 90–3.

Garfield, E. and Sher, I.H. (1963) 'New factors in the evaluation of scientific literature through citation indexing', *American Documentation*, 14(3): 195–201.

Garg, K.C. (2002) 'Scientometrics of laser research in India and China', *Scientometrics*, 55(1): 71–85.

Giles, J. (2005) 'Start your engines', *Nature*, 438(7068): 554–5.

Glänzel, W. (2003) *Bibliometrics as a Research Field. A Course on Theory and Application of Bibliometric Indicators*. Course handouts. Online at: *http://scholar.google.com*.

Glänzel, W. and Schoepflin, U. (1995) 'A bibliometric study on ageing and reception processes of scientific literature', *Journal of Information Science*, 21: 37–53.

Glänzel, W. and Schubert, A. (2001) 'Double effort = double impact? A critical view at international co-authorship in chemistry', *Scientometrics*, 50(2): 199–214.

Glänzel, W., Thus, B. and Schlemmer, B. (2004) 'A bibliometric approach to the role of author self-citations in scientific communication', *Scientometrics*, 59(1): 63–77.

Gmür, M. (2003) 'Co-citation analysis and the search for invisible colleges: a methodological evaluation', *Scientometrics*, 57(1): 27–57.

Gómez-Benito, J., Hidalgo-Montesinos, M.D., Guilera-Ferré, G. and Moreno-Torrente, M. (2005) 'A bibliometric study of differential item functioning', *Scientometrics*, 64(1): 3–16.

González-Alcaide, G., Aleixandre-Benavent, R., Navarro-Molina, C. and Valderrama-Zurián, J.C. (2008) 'Coauthorship networks and institutional collaboration patterns in reproductive biology', *Fertility and Sterility*, 90(4): 941–56.

Gorkova, V.I. (1988) 'Informetrics', *Informatics, 10*. Moscow: VINITI.

Green, J.B. (2008) 'Limiting the impact of the impact factor', *Science*, 322: 1463.

Gregory, J. (1937) 'An evaluation of medical periodicals', *Bulletin of the Medical Library Association*, 25(3): 172–88.

Gross, O.V. (1967) 'Bradford's law and the Keenen-Atherton data', *American Documentation*, 18(1): 46.

Gross, P.L.K. and Gross, E.M. (1927) 'College libraries and chemical education', *Science*, 66(1713): 385–9.

Guan, J. and Ma, N. (2004) 'A comparative study of research performance in computer science', *Scientometrics*, 61(3): 339–59.

Guan, J. and Ma, N. (2007) 'A bibliometric study of China's semiconductor literature compared with other major Asian countries', *Scientometrics*, 70(1): 107–24.

Guns, R. and Rousseau, R. (2009) 'Real and rational variants of the h-index and the g-index', *Journal of Informetrics*, 3: 64–71.

Habibzadeh, F. and Yadollahie, M. (2008) 'Journal weighted impact factor: a proposal', *Journal of Informetrics*, 2: 164–72.

Hamilton, D. (1991) 'Research papers: who's uncited now?', *Science*, 251: 25.

Hanley, L.M. (1937) *Word Index to James Joyce's Ulysses*. Madison, WI: University of Wisconsin Press.

Haynes, B. (2007) 'Reliance on bibliometric databases can let you down', *Nature*, 466: 725.

Heine, M. (1998) 'Bradford ranking conventions and their application to a growing literature', *Journal of Documentation*, 54(3): 303–31.

Hicks, D. (2004) 'The four literatures of the social science', in H.K. Moed, W. Glänzel and U. Schmoch (eds), *Handbook of Quantitative Science and Technology Research*. Dordrecht: Springer.

Hirsch, J.E. (2005) 'An index to quantify an individual's scientific research output', *Proceedings of the National Academy of Sciences of the United States of America*, 15: 16569–72.

Hirsch, J.E. (2007) 'Does the h index have predictive power?', *Proceedings of the National Academy of Sciences*, 104: 19193–8.

Holden, C. (2005) 'Random samples: data point – impact factor', *Science*, 309: 1181.

Hönekopp, J. and Kleber, J. (2008) 'Sometimes the impact factor outshines the H index', *Retrovirology*, 5: 88–90.

Hood, W.W. and Wilson, C.S. (2003) 'Informetric studies using databases: opportunities and challenges', *Scientometrics*, 58(3): 587–608.

Hou, H., Kretschmer, H. and Liu, Z. (2008) 'The structure of scientific collaboration networks in Scientometrics', *Scientometrics*, 75(2): 189–202.

Hubert, J.J. (1981) 'General bibliometric models', *Library Trends*, 30(1): 65–81.

Inzelt, A., Schubert, A. and Schubert, M. (2009) 'Incremental citation impact due to international co-authorship in Hungarian higher education institutions', *Scientometrics*, 78(1): 37–43.

Jacso, P. (2001) 'A deficiency in the algorithm for calculating the impact factor of scholarly journals: the journal impact factor', *Cortex*, 37(4): 590–4.

Jacso, P. (2005) 'As we may search – comparison of major features of Web of Science, Scopus and Google Scholar citation-based and citation-enhanced databases', *Current Science*, 89(9): 1537–47.

Jacso, P. (2006) 'Deflated, inflated and phantom citation counts', *Online Information Review*, 30(3): 297–309.

Jacso, P. (2008) 'The pros and cons of computing the h-index using Scopus', *Online Information Review*, 32(4): 524–35.

Jeang, K.T. (2007) 'Impact factor, H index, peer comparisons, and retrovirology: is it time to individualize citation metrics?', *Retrovirology*, 4: 42.

Jemec, G.B. (2001) 'Impact factors of dermatological journals for 1991–2000', *BMC Dermatology*, 1: 7.

Jin, B.H. (2006) 'H-index: an evaluation indicator proposed by scientist', *Science Focus*, 1(1): 8–9 (in Chinese).

Jin, B.H. (2007) 'The AR-index: complementing the h-index', *ISSI Newsletter*, 3(1): 6.

Jin, B.H., Liang, L., Rousseau, R. and Egghe, L. (2007) 'The R- and AR-indices: complementing the h-index', *Chinese Science Bulletin*, 52(6): 855–63.

Joseph, K.S. and Hoey, J. (1999) 'CMAJ's impact factor: room for recalculation', *Canadian Medical Association Journal*, 161(8): 977–8.

Kelly, C.D. and Jennions, M.D. (2006) 'The h index and career assessment by numbers', *Trends in Ecology and Evolution*, 21(4): 167–70.

Kendall, M.G. (1960) 'The bibliography of operational research', *Operational Research Quarterly*, 11: 31–6.

Kirchler, E. and Hölzl, E. (2006) 'Twenty-five years of the *Journal of Economic Psychology* (1981–2005): a report on the development of an interdisciplinary field of research', *Journal of Economic Psychology*, 27: 793–804.

Kochen, M. (1974) *Principles of Information Retrieval*. New York: Wiley.

Kosmulski, M. (2006) 'A new Hirsch-type index saves time and works equally well as the original h-index', *ISSI Newsletter*, 2(3): 4–6.

Kretschmer, H. (2004) 'Author productivity and geodesic distance in bibliographic co-authorship networks, and visibility on the Web', *Scientometrics*, 60(3): 409–20.

Kreuzman, H. (2001) 'A co-citation analysis of representative authors in philosophy: examining the relationship between epistemologists and philosopers of science', *Scientometrics*, 51(3): 525–39.

Ladwig, J.P. and Sommese, A.J. (2005) 'Using cited half-life to adjust download statistics', *College and Research Libraries*, 66(6): 527–42.

LaGuardia, C. (2005) 'E-views and reviews: Scopus vs. Web of Science', *Library Journal*, 130(1): 40–2.

Lascar, C. and Mendelsohn, L.D. (2001) 'An analysis of journal use by structural biologists with application for journal collection development decisions', *College and Research Libraries*, 62(5): 422–33.

Leimkuhler, F.F. (1967) 'The Bradford distribution', *Journal of Documentation*, 23: 197–207.

Levine-Clark, M. and Gil, E.L. (2009) 'A comparative citation analysis of Web of Science, Scopus and Google Scholar', *Journal of Business and Finance Librarianship*, 14(1): 32–46.

Lindsey, D. (1980) 'Production and citation measures in the sociology of science: the problem of multiple authorship', *Social Studies of Science*, 10: 145–62.

Liu, Y. and Rousseau, R. (2007) 'Hirsch-type indices and library management: the case of Tongji University library', *Proceedings of the ISSI 2007: 11th International Conference of the International Society for Scientometrics and Informetrics*, 2 vols. Madrid: CINDOC-CSIC, pp. 514–22.

Liu, Z. (2005) 'Visualizing the intellectual structure in urban studies: a journal co-citation analysis (1992–2002)', *Scientometrics*, 62(3): 385–402.

López-Muñoz, F., Vieta, E., Rubio, G., García-García, P. and Alamo, C. (2006) 'Bipolar disorder as an emerging pathology in the scientific literature: a bibliometric approach', *Journal of Affective Disorders*, 92: 161–70.

Lotka, A.J. (1926) 'The frequency distribution of scientific productivity', *Journal of the Washington Academy of Sciences*, 16(12): 317–23.

Luwel, M. and Moed, H.F. (1998) 'Publication delays in the science field and their relationship to the aging of scientific literature', *Scientometrics*, 41(1–2): 26–40.

McCain, K.W. (1990) 'Mapping authors in intellectual space: a technical overview', *Journal of the American Society for Information Science*, 41: 433–43.

McDonald, J.D. (2007) 'Understanding journal usage: a statistical analysis of citation and use', *Journal of the American Society for Information Science and Technology*, 58(1): 39–50.

McNeely, J.K. and Crosno, C.D. (1930) 'Periodicals for electrical engineers', *Science*, 72(1856): 81–4.

McVeigh, M.E. (2004) *Open Access Journals in the ISI Citation Databases: Analysis of Impact Factors and Citation Patterns*. Philadelphia, PA: Thomson Corporation.

Magyar, G. (1974) 'Bibliometric analysis of a new research sub-field', *Journal of Documentation*, 30(1): 32–40.

Marchi, M.D. and Rocchi, M. (2001) 'The editorial policies of scientific journals: testing an impact factor model', *Scientometrics*, 51(2): 395–404.

Martin, B.R. (1996) 'The use of multiple indicators in the assessment of the basic research', *Scientometrics*, 36(3): 343–62.

Marton, J. (1985) 'Obsolence for immediacy? Evidence supporting Price's hypothesis', *Scientometrics*, 7: 145–53.

Martyn, J. and Gilchrist, A. (1968) *An Evaluation of Scientific Journals*, Aslib Occasional Publication No. 1. London: Aslib.

Meho, L.I. and Yang, K. (2007) 'Impact of data sources on citation counts and rankings of LIS faculty: Web of Science versus Scopus and Google Scholar', *Journal of the American Society for Information Science and Technology*, 58(13): 2105–25.

Merton, R.K. (1968) *Social Theory and Social Structure*. New York: Free Press.

Miguel-Dasit, A., Martí-Bonmatí, L. and Sanfeliu, P. (2008) 'Bibliometric analysis of the Spanish MR radiological production (2001–2007)', *European Journal of Radiology*, 67: 384–91.

Moed, H.F. (2002a) 'Measuring China's research performance using the Science Citation Index', *Scientometrics*, 53(3): 281–96.

Moed, H.F. (2002b) 'The impact factors debate: the ISI's uses and limits', *Nature*, 415: 731–2.

Moed, H.F. and van Leeuwen, T.N. (1995) 'Improving the accuracy of the Institute for Scientific Information's Journal Impact factors', *Journal of the American Society for Information Science*, 46: 461–7.

Moed, H.F., van Leeuwen, T.N. and Reedijk, J. (1998) 'A new classification system to describe the ageing of scientific journals and their impact factors', *Journal of Documentation*, 54: 387–419.

Moed, H.F., van Leeuwen, T.N. and Reedijk, J. (1999) 'Toward appropriate indicators of journal impact', *Scientometrics*, 46(3): 575–89.

Moed, H.F., Burger, W.J.M., Frankfort, J.G. and van Raan, A.F.J. (1985) 'The application of bibliometric indicators: important field and time-dependent factors to be considered', *Scientometrics*, 8: 177–203.

Monastersky, R. (2005) 'The number that's devouring science', *Chronicle of Higher Education*, 52(8): A12.

Mullen, L.B. and Hartman, K.A. (2006) 'Google Scholar and the Library Web Site: the early response by ARL Libraries', *College and Research Libraries*, 67(2): 106–22

Nalimov, V.V. and Mulchenko, B.M. (1969) *Scientometrics*. Moscow: Nauca.

Narin, F. (1976) *Evaluative Bibliometrics: The Use of Publication and Citation Analysis in the Evaluation of Scientific Activity*. Cherry Hill, NJ: Computer Horizons.

Narin, F., Stevens, K., and Whitlow, E.S. (1991) 'Scientific cooperation in Europe and the citation of multinationally authored papers', *Scientometrics*, 21(3): 313–23.

Nederhof, A. (2006) 'Bibliometric monitoring of research performance in the social sciences and the humanities: a review', *Scientometrics*, 66(1): 81–100.

Neuhaus, C. and Daniel, H.D. (2008) 'Data sources for performing citation analysis: an overview', *Journal of Documentation*, 64(2): 193–210.

Nicholls, P.T. (1986) 'Empirical validation of Lotka's law', *Information Processing and Management*, 22(5): 417–19.

Norris, M. and Oppenheim, C. (2007) 'Comparing alternatives to the Web of Science for coverage of the social sciences literature', *Journal of Informetrics*, 1: 161–9.

Pao, M.L. (1985) 'Lotka's law: a testing procedure', *Information Processing and Management*, 21(4): 305–20.

Pauly, D. and Stergiou, K.I. (2005) 'Equivalence of results from two citation analyses: Thomson's ISI Citation Index and Google's Scholar service', *Ethics in Science and Environmental Politics*, 22 December, pp. 33–5.

Persson, O. (2001) 'All author citations versus first author citations', *Scientometrics*, 50(2): 339–44.

Pinski, G. and Narin, F. (1976) 'Citation influence for journal aggregates of scientific publications: theory, with application to literature of physics', *Information Processing and Management*, 12(5): 297–312.

Posner, R.A. (2001) *Public Intellectual: A Study of Decline*. Cambridge, MA: Harvard University Press.

Price, D.J. (1963) *Little Science, Big Science*. New York: Columbia University Press.

Pritchard, A. (1969) 'Statistical bibliography or bibliometrics?', *Journal of Documentation*, 24: 348–9.

Pudovkin, A.I. and Garfield, E. (2002) 'Algorithmic procedure for finding semantically related journals', *Journal of the American Society for Information Science and Technology*, 53: 1113–19.

Pulgarín, A. and Gil-Leiva, I. (2004) 'Bibliometric analysis of the automatic indexing literature: 1956–2000', *Information Processing and Management*, 40: 365–77.

Rahm, E. and Thor, A. (2005) 'Citation analysis of database publication', *SIGMOD Record*, 34(4): 48–53.

Redner, S. (2005) 'Citation statistics from 110 years of *Physical Review*', *Physics Today*, 58: 49–56.

Ren, S. and Rousseau, R. (2002) 'A scientometric data analysis of JCR-covered journals in geo-sciences', *Journal of Physics*, 28(1): 4–13.

Roediger, H.L. (2006) 'The h-index in science: a new measure of scholarly contribution', *Academic Observer*, 19(4). Online at: *http://www.psychologicalscience.org/observer/19/4/academic-observer*.

Rousseau, R. (1988) 'Citation distribution of pure mathematics journals', in L. Egghe and R. Rousseau (eds), *Informetrics 87/88*. Amsterdam: Elsevier Science.

Rousseau, R. (1992) 'Breakdown of the robustness property of Lotka's law: the case of adjusted counts for multiauthorship attribution', *Journal of the American Society for Information Science*, 43(10): 645–7.

Rousseau, R. (1994) 'Bradford curves', *Information Processing and Management*, 30: 267–77.

Rousseau, R. (2008) 'Woeginger's axiomatisation of the h-index and its relation to the g-index, the h(2)-index and the R^2-index', *Journal of Informetrics*, 2: 335–40.

Rousseau, R. and Leimkuhler, F.F. (1987) 'The nuclear zone of a Leimkuhler curve', *Journal of Documentation*, 43(4): 322–33.

Rousseau, R. and van Hooydonk, G. (1996) 'Journal production and journal impact factors', *Journal of the American Society for Information Science*, 47(10): 775–80.

Saad, G. (2006) 'Exploring the h-index at the author and journal levels using bibliometric data of productive consumer scholars and business-related journals respectively', *Scientometrics*, 69(1): 117–20.

Schreiber, M. (2008) 'The influence of self-citation corrections on Egghe's g index', *Scientometrics*, 76(1): 187–200.

Schubert, A. (2007) 'Successive h-indices', *Scientometrics*, 70(1): 201–5.

Schubert, A. and Braun, T. (1986) 'Relative indicators and relational charts for comparative assessment of publication output and citation impact', *Scientometrics*, 9: 281–91.

Scott, J.P. (1991) *Social Network Analysis: A Handbook*. London: Sage.

Simons, K. (2008) 'The misused impact factor', *Science*, 322: 165.

Simonton, D.K. (1997) 'Creative productivity: a predictive and explanatory model of carrier trajectories and landmarks', *Psychological Review*, 104: 66–89.

Simonton, D.K. (2003) 'Scientific creativity as constrained stochastic behaviour: the integration of product, person, and process perspectives', *Psychological Bulletin*, 129: 475–94.

Snyder, H. and Bonzi, S. (1998) 'Patterns of self-citation across disciplines (1980–1989)', *Journal of Information Science*, 24(6): 431–5.

Su, Y.M., Yang, S.C., Hsu, P.Y. and Shiau, W.L. (2009) 'Extending co-citation analysis to discover authors with multiple expertise', *Expert Systems with Applications*, 36: 4287–95.

Thomson Reuters (1994) *The Thomson Scientific Impact Factor*. Current Contents from the Institute for Scientific Information. Online at: *http://www.thomsonreuters.com/products_services/science/free/essays/impact_factor*.

Thomson Reuters (2009) *Journal Citation Reports® – 'The JCR' February 2009 Enhancements*. Thomson Reuters. Online at: *http://www.accesowok.fecyt.es/news/2009/JCR-Feb2009Enhancements2.ppt*.

Tol, R.S.J. (2008) 'A rational, successive g-index applied to economics departments in Ireland', *Journal of Informetrics*, 2: 149–55.

Tsay, M.Y. (2009) 'An analysis and comparison of scientometric data between journals of physics, chemistry and engineering', *Scientometrics*, 78(2): 279–93.

Tsay, M.Y., Jou, S.J. and Ma, S.S. (2000) 'A bibliometric study of semiconductor literature, 1978–1997', *Scientometrics*, 49(3): 491–509.

van Raan, A.F.J. (1996) 'Advanced bibliometric methods as quantitative core of peer review based evaluation and foresight exercises', *Scientometrics*, 36: 397–420.

van Raan, A.F.J. (2000) 'The Pandora's box of citation analysis: measuring scientific excellence, the last evil?', in C. Cronin and H.D. Atkins (eds), *The Web of Knowledge. A Festschrift in Honor of Eugene Garfield*. Medford, NJ: ASIS Monograph Series.

van Raan, A.F.J. (2004) 'Measuring science. Capita selecta of current main issues', in H.F. Moed, W. Glänzel and U. Schmoch (eds), *Handbook on Quantitative Science and Technology Research*. Dordrecht: Kluwer Academic.

van Raan, A.F.J. (2005) 'Fatal attraction: conceptual and methodological problems in the ranking of universities by bibliometric methods', *Scientometrics*, 62(1): 133–43.

van Raan, A.F.J. (2006a) 'Performance-related differences of bibliometric statistical properties of research groups: cumulative advantages and hierarchically layered networks', *Journal of the American Society for Information Science and Technology*, 57(14): 1919–35.

van Raan, A.F.J. (2006b) 'Comparisons of the Hirsch-index with standard bibliometric indicators and with peer judgment for 147 chemistry research groups', *Scientometrics*, 67(3): 491–502.

van Raan, A.F.J. (2008) 'Self-citation as an impact-reinforcing mechanism in the science system', *Journal of the American Society for Information Science and Technology*, 59(10): 1631–43.

van Raan, A.F.J. and van Leeuwen, T.N. (2002) 'Assessment of the scientific basis of interdisciplinary, applied research. Application of bibliometric methods in nutrition and food research', *Research Policy*, 31: 611–32.

Vanclay, J.K. (2009) 'Bias in the journal impact factor', *Scientometrics*, 78(1): 3–12.

Vinkler, P. (2004) 'Characterization of the impact of sets of scientific papers: the Garfield (Impact) Factor', *Journal of the American Society for Information Science and Technology*, 55: 431–5.

Wallace, D.P. (1985) *The Relationship Between Journal Productivity and Obsolence in a Subject Literature*. PhD dissertation, University of Illinois at Urbana-Champaign.

Wasserman, S. and Faust, K. (1994) *Social Network Analysis*. Cambridge: Cambridge University Press.

Weinberg, A. (1961) 'Impact of large-scale science on the US', *Science*, 134: 161–4.

Weingart, P. (2005) 'Impact of bibliometrics upon the science system: inadvertent consequences?', *Scientometrics*, 62(1): 117–31.

Westbrook, J.H. (1960) 'Identifying significant research', *Science*, 132(3435): 1229–34.

White, H.D. and Griffith, B.C. (1981) 'Author cocitation: a literature measure of intellectual structure', *Journal of the American Society for Information Science*, 32(3): 163–71.

Wilkinson, E.A. (1973) *The Bradford-Zipf Distribution. A Simulation Study*. School of Library, Archive and Information Studies, University College, London.

Willett, P. (2007) 'A bibliometric analysis of the *Journal of Molecular Graphics and Modelling*', *Journal of Molecular Graphics and Modelling*, 26: 602–6.

Wilson, C.S. (1999) 'Informetrics', in M. Williams (ed.), *Annual Review of Information Science and Technology*. Medford, NJ: Information Today.

Woeginger, G.J. (2008) 'An axiomatic analysis of Egghe's g-index', *Journal of Informetrics*, 2: 364–8.

Wolfram, D. (2003) *Applied Informetrics for Information Retrieval Research*. Westport, CT: Libraries Unlimited.

Xie, S., Zhang, J. and Ho, Y.S. (2008) 'Assessment of world aerosol research trends by bibliometric analysis', *Scientometrics*, 77(1): 113–30.

Yancey, R. (2005) *Fifty Years of Citation Indexing and Analysis*. Newsletter from Thomson Scientific.

Yu, G., Go, R. and Li, Y.J. (2006) 'The influence of publication delays on three ISI indicators', *Scientometrics*, 69(3): 511–27.

Yu, G., Wang, X.H. and Yu, D.R. (2005) 'The influence of publication delays on impact factors in the scientific field', *Scientometrics*, 64(2): 235–46.

Zhivotovsky, L.A. and Krutovsky, K.V. (2008) 'Self-citation can inflate h-index', *Scientometrics*, 77(2): 373–5.

Zipf, G.K. (1935) *The Psycho-biology of Language*. Cambridge, MA: MIT Press.

Zipf, G.K. (1949) *Human Behavior and the Principle of Least Effort*. Cambridge, MA: Addison-Wesley.

Index

Kolibri 360 GmbH
Ferdinand-Jühlke-Straße 7
99095 Erfurt, Deutschland
produktsicherheit@kolibri360.de